Opioid Food Peptides

Mohammad Raies Ul Haq

Opioid Food Peptides

Significant Exorphins from Food Sources

Mohammad Raies Ul Haq
Department of Biochemistry,
Sri Pratap College
Cluster University Srinagar
Srinagar, Jammu and Kashmir, India

ISBN 978-981-15-6104-7   ISBN 978-981-15-6102-3   (eBook)
https://doi.org/10.1007/978-981-15-6102-3

This Springer imprint is published by the registered company Springer Nature Singapore Pte Ltd.
The registered company address is: 152 Beach Road, #21-01/04 Gateway East, Singapore 189721,
Singapore

• My parents, respected Mohammad Yousuf Mir and Amina Banoo, for their consistent support
• My wife, Urusha Raies, for bearing with me in all respects
• My sister, Towheeda Yousuf
• Dr. Arshid Hussain
• My kids, Juwariya Batool, Tohoora Hisaan, and Mohammad Faaiz-Ul-Haq

Preface

This book entitled *"Opioid Food Peptides: Significant Exorphins from Food Sources"* describes vividly the production of exorphins from daily foods like wheat, rice, milk, soybean, and spinach. The exorphins include casomorphins, lactorphins, lactoferroxins, and casoxins (milk), gluten exorphins (wheat), rubiscolins (spinach RuBisCo), soymorphins (soybean), and oryzatensin (rice albumin). The structural features of these exorphins are described that are critical for the demonstration of opioid activity. These peptides have structural resemblance with the endorphins and morphine and thereby bind endogenous opioid receptors. This book explains in detail the opioid activity of these exorphins assessed through pharmacological techniques. Moreover, a vibrant description of the classification of these exorphins is given in each chapter. Moreover, the correlations between increased production and absorption of these peptides and manifestations of human complications are also given in detail. The beneficial roles of these peptides are highlighted including roles in gastrointestinal motility, analgesia, anxiolysis, emotional and behavior development, memory consolidation, blood pressure regulation, prolactin secretion, food and fat intake, hormone release, appetite, and mucous formation. This book demonstrates the advantages associated with food-derived exorphins over conventional opioids. These merits include the least side effects, present oral bioavailability, and activate endogenous opioid receptor systems. This book highlights that these exorphins may have enough scope in the food and pharmaceutical industry, besides biomedical sciences for the development of therapeutic mediators, functional foods, and nutraceuticals. However, verified and authenticated research with reproducible results is needed in this field. This book is valuable for the students and researchers of agricultural and animal sciences, nutritional and biomedical sciences, as well as industry experts.

Srinagar, India   Mohammad Raies Ul Haq

Acknowledgments

Searching literature and then working in the field of food exorphins was an awesome but challenging experience. The practical and therotical experience was expolited in the composition of this book, largely during the COVID-19 lockdown period.

In this endeavor, first and foremost I thank Almighty Allah who retained me in circumstances more favorable than acquiring this successful attempt. He (Allah) bestowed upon me with the lovely family (Dad, Mom, Urusha, Sister, Dr. Arshid, Juwariya, Tahoora, and Faaiz) as a sounding board at both favorable and critical times.

I acknowledge Mr. Khursheed, Dr. Shafi, Dr. Gulzar, Dr. Muzamil, Dr. Iqbal, Dr. Shabir, Dr. Showkeen, Dr. Sheraz, Dr. Danish, and Dr. Ahmad for the entertainment they provided that was subsequently essential for sustenance of balanced and sound mind.

The composition of this book would have been impossible without some exceptional research work by many eminent scientists who made the discoveries on which this book is based. One or two such scientists whom I corresponded with and the remaining are known to me through their outstanding research.

Through the medium of this book, I did not try to make any nutritional recommendation but it is a collection of literature and my research expertise.

I tried to write this book with the hope that it is free of scientific errors. Nevertheless, it is just a hope, science is occasionally free of errors and omissions. The development of a scientific base is through repudiation of the principles that orthodox perception articulates are correct. The errors in the book are my responsibility.

I welcome the suggestions and critical analysis for the next edition of this book.

Contents

About the Author

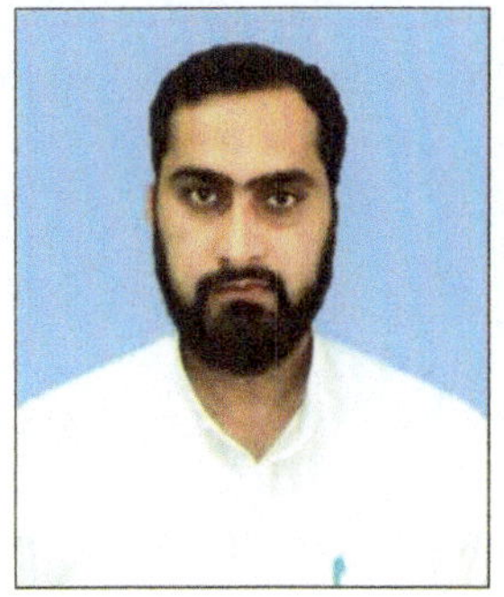

Mohammad Raies Ul Haq holds a Ph.D. in Biochemistry from the National Dairy Research Institute-Indian Council of Agricultural Research (NDRI-ICAR), Karnal, Haryana. He is currently an Assistant Professor of Biochemistry at Sri Pratap College, Srinagar. He is an Editorial Board Member of the Journal of Dentistry & Oral Disorders and of Acta Scientific Nutritional Health. He is a member of The Society of Biological Chemists and the Indian Science Congress and a member of the Board of Undergraduate and Postgraduate Studies for Biochemistry and Clinical Biochemistry at the Cluster University of Srinagar. He has more than 5 years of teaching experience at various government colleges and universities. He worked as Lecturer at the Department of Biochemistry, University of Kashmir, and at the Government College for Women, M. A. Road, Srinagar. He also passed the National Eligibility Test (NET) conducted by CSIR. He has authored three books: "Analytical Biochemistry," "Skill Development in Biochemistry," and "Expertise Enhancement in Biochemistry" and published several research papers in reputed international journals. He was awarded a "Certificate of Appreciation" from NDRI-ICAR for his outstanding Ph.D. research. Moreover, he was a Postdoctoral Fellow at the Center of Research for Development (CORD) at University of Kashmir (UoK), where he conducted research funded by SERB-DST, New Delhi.

Abbreviations

ACD	Asymptomatic celiac diseases
ACE	Angiotensin-I-converting enzyme
ACN	Acetonitrile
ACRS	Amplification created restriction site
ALTE	Acute life-threatening events
ASDs	Autistic spectrum disorders
AS-PCR	Allele specific-polymerase chain reaction
BAP	Bioactive peptides
BB	BioBreeding
BBB	Blood–brain barrier
BCM	β-Casomorphin
BCM-5	β-Casomorphin-5
BCM-7	β-Casomorphin-7
BCMIR	β-Casomorphin immunoreactive
BSA	Bovine serum albumin
CD	Cluster of differentiation
CD	Celiac diseases
CHD	Coronary heart disease
COX	Cyclooxygenase
CSF	Cerebrospinal fluid
CVD	Cardiovascular disease
DIPFP	Diisopropylfluorophosphate
DOP	Delta-opioid receptor
DPP IV	Dipeptidyl peptidase IV
ED	Effective dose
EFSA	European Food Safety Authority
ELISA	Enzyme-linked immunosorbent assay
EMC	Enzyme-modified Cheddar cheese
FAO	Food and Agriculture Organization
FFDCA	The U.S. Federal Food, Drug, and Cosmetic Act
GABA	Gamma-aminobutyric acid
GER	Gastric emptying rate
GIP	Gastrin inhibitory peptide

GITT	Gastrointestinal transit time
GLUT-2	Glucose transporter 2
GPCRs	G protein-coupled receptors
GPI	Guinea pig ileum
HDL	High-density lipoprotein
HLA	Human leukocyte antigen
ICAR	Indian Council of Agricultural Research
ICV	Intracerebroventricular
IEC	Isoelectric casein
IF	Infant formula
IHD	Ischemic heart disease
ISI	IdeaSphere Incorporated
IUPHAR	International Union of Basic and Clinical Pharmacology
KF	Karan Fries
KOP	Kappa-opioid receptor
KS	Karan Swiss
LAB	Lactic acid bacteria
LAP	Leucine aminopeptidase
LDL	Low-density lipoprotein
LPL	Lamina propria lymphocytes
MALDI	Matrix-assisted laser desorption/ionization
MALDI-TOF	Matrix-assisted laser desorption/ionization-time of flight
MCP-1	Monocyte chemotactic protein-1
MMP	Matrix metalloproteinase
MMR	Measles–mumps–rubella
MOP	Mu-opioid receptor
MPO	Myeloperoxidase
MS	Mass spectroscopy
MUC2	Mucin 2
MVD	Mouse vas deferens
NDRI	National Dairy Research Institute
NO	Nitric oxide
NOD	Non-obese diabetic mice
NOS	Nitric oxide synthase
ODS	Octadecyl silica
oxLDL	Oxidized low-density lipoprotein
PBMCs	Peripheral blood mononuclear cells
PBS	Phosphate buffer saline
PGD	Prostaglandin
PO	Prolyl oligopeptidase
RANKL	Nuclear factor kappa-B ligand
RFLP	Restriction fragment length polymorphism
ROS	Reactive oxygen species
RP-HPLC	Reversed-phase high-performance liquid chromatography

RR	Relative risk
RT	Retention time
RuBisCo	D-ribulose-1,5-bisphosphate carboxylase/oxygenase
SCFA	Short-chain fatty acids
SCIT	Subtle cognitive impairment test
SDS-PAGE	Sodium dodecyl sulphate-polyacrylamide gel electrophoresis
SGID	Simulated gastrointestinal digestion
SHR	Spontaneously hypertensive rats
SIDS	Sudden infant death syndrome
SSC	Somatic cell count
SSCP	Single-strand conformation polymorphism
T1D	Type 1 diabetes mellitus
TEA	Tetraethylammonium
TLR	Toll-like receptor
TOF	Time of flight
UHT	Ultra heat treatment
USDA	United States Department of Agriculture
VEGF	Vascular endothelial growth factor
VLDL	Very low-density lipoprotein
WKY	Wistar-Kyoto

Significant Food-Derived Opioid Peptides

Abstract

Bioactive peptides (BAPs) are biologically active protein fragments. BAPs are released from the intact food proteins during gastrointestinal digestion, fermentation, bacterial cultures, or a combination of these processes. Food-derived exorphins are BAPs released from foods that show opioid activity. These include casomorphins, lactorphins, lactoferroxins, and casoxins (milk), gluten exorphins (wheat), rubiscolins (spinach), soymorphins (soybean), and oryzatensin (rice). They have selectivity for the endogenous opioid receptors including mu (μ), delta (δ), and kappa (κ). These opioid receptors are distributed in central and peripheral nervous systems, gastrointestinal tract, immune cells, and some other tissues. These exorphins possess all the structural features that are critical for binding these opioid receptors and in turn manifestation of opioid activity. They show different physiological activities like analgesic property, anxiolysis, hormone secretion, gastrointestinal transit, ACE inhibition, lymphocyte proliferation, food and fat intake, memory consolidation, and behavioral and pharmacological effects. Moreover, the consumption of a few exorphins is correlated with some human illnesses. Overall, these peptides have enough scope in the food and pharmaceutical industry for the development of functional foods and nutraceuticals.

1.1 Introduction

Bioactive peptides (BAPs) are physiologically active fragments derived from the protein family. BAPs from food sources are short stretches of amino acid residues linked through peptide bonds (Shahidi and Zhong 2008; Sharma et al. 2011; Walther and Sieber 2011). These are organic substances mostly encrypted in the long protein biomolecules, although, a very few of them may exist in free form. These peptide fragments may be released from the intact protein molecules during gastrointestinal digestion, fermentation, bacterial cultures, or a combination of these processes (Korhonen and Pihlanto 2003, 2006; Korhonen 2009). These peptides may bind with various receptors in the central and peripheral nervous system, gastrointestinal tract, and some immune cells. The peptides act as the ligands for these receptors and therefore regulate various physiological effects in humans. There are many reports available that demonstrate the beneficial role of these fragments in the human physiology (Kitts and Weiler 2003). More recently, the beneficial attributes have been exploited by numerous health professionals for the development of novel dietary additives and functional products (Sánchez and Vázquez 2017). These peptides play role in the intermediary metabolism and therefore demonstrate close associations with the gastrointestinal tract, cardiovascular system, immune functions, and nervous system in humans. The analytical approach to decipher peptide quantification is significant for health professionals in order to elucidate their functional role in human nutrition, health, and disease (Mora et al. 2014). Approximately 1500 different BAPs are available in the peptide database and entitled as "Biopep" (Singh et al. 2014). Many of these BAPs establish hormone or drug "like" functions and are further categorized as exorphins (opioids), anxio-lytic, antimicrobial, antihypertensive, antithrombotic, immunomodulatory, antioxidative, and mineral binding. The composition and sequence of amino acids in the BAP determines their functional attributes. The metabolic and immune functions may alter BAP-receptor binding in the human body (Fields et al. 2009). Plant and animal proteins are a potential source of diverse BAPs when produced inside the gastrointestinal tract by proteolytic enzymes (Carrasco-Castilla et al. 2012; Bhat et al. 2015). Although the structural-functional relationship of the BAPs is not fully explored, however, few attributes have been demonstrated in this perspective. It is established that food-derived BAPs are composed of 2–20 amino acids. Moreover, they may show hydrophobic in addition to basic amino acid residues (lysine or arginine) and proline. The presence of proline makes them resistant to various proteolytic enzymes except for prolyl proteases (dipeptidyl-peptidase IV) expressed by brush border in the gut (Moller et al. 2008; Kitts and Weiler 2003). Currently, their activities may be exploited in overcoming few adverse functions like oxidative stress and microbial pathogenesis. Therefore they have potential in ameliorating the quality of life (Lemes et al. 2016). More recently, in the food and pharmaceutical industry, the use of nutraceuticals and functional foods has gained much consider-ation as the potential impact they may demonstrate in human nutrition, health, and disease (Haque et al. 2008; Moldes et al. 2017). Subsequently, from the last few decades, special attention has been given to the generation, properties, and functions of BAPs (Przybylski et al. 2016). On one hand, the growing health professionals are hopeful for the potential beneficial role of these peptides in various human complications. On the other hand, some epidemiological data has been generated that correlate consumption of food-derived opioid peptides with an increased risk for various health complications including type 1 diabetes (Elliott et al. 1999; McLachlan 2001; Virtanen et al. 2000; Laugesen and Elliott 2003a, b; Birgisdottir et al. 2002; Monetini et al. 2002), heart diseases (McLachlan 2001; Laugesen and Elliott 2003a, b; Tailford et al. 2003), neurological manifestations (Sun et al. 1999,

2003, 1999; Cade et al. 2000; Lázaro et al. 2016; Bojović et al. 2017), and gut inflammation (Raies et al. 2014a, b).

1.2   Basic Structural Properties of Exorphins

The opioid peptides derived from the foods or microbes are known as exorphins. These are produced from the proteins of the foods on hydrolysis through the proteolytic system. They mimic the endogenous opioids as they have the potential to bind the endogenous opioid receptors like that of endorphins. The common food-derived opioid peptides are as follows:

- Casomorphins derived from α and β-casein of human and bovine milk.
- Casoxins derived from the κ-casein of bovine milk.
- Lactorphins released from α-lactalbumin and β-lactoglobulin, respectively.
- Lactoferroxin derived from bovine milk lactoferrin.
- Gluten exorphins released from the gluten of wheat, barley, rye, etc.
- Rubiscolins obtained from spinach.
- Soymorphins produced from soybean.
- Oryzatensin obtained from rice albumin.

It is very interesting to note that morphine and β-casomorphins (BCMs) belong to two different categories. Morphine is an alkaloid while BCMs are opioid peptides. Despite the class dissimilarity, these molecules bind with the same endogenous opioid receptors in humans. The possible explanation is the similarity in the two molecules they share in their structure. There is also a logical dependence of a biological action on the structural attribute. The biological response should not only rely on the steric configuration but on the electronic properties as well. It implies that a thermodynamically stable state may persist in every active molecule that owns the possibility of a comparable interaction with the structurally unidentified receptor (Brandt et al. 1994). Morphine has been found to have a high affinity for the μ-receptors, BCMs also demonstrate an affinity for the same receptor, although the affinity is low (Liebmann et al. 1986). Brandt and coworkers in 1994 established that the presence of nitrogen atom in the morphine in a well-defined spatial arrangement with respect to the aromatic ring (A) confers the structural features needed for the demonstration of opioid activity. Further, the existence of a phenolic hydroxyl group helps in strengthening the ligand-receptor affinity. Nevertheless, the latter is not essential for the manifestation of opioid activity. Moreover, the presence of the second benzene ring (F-ring) also appears significantly important (Brandt et al. 1994). It is of remarkable interest to highlight that these structural elements also exist in milk-derived BCMs. Therefore, these structural resemblances provide enough probabilities for these two different classes of compounds to demonstrate similar opioid activities. Food-derived opioid peptides are also regarded as "formones or food hormones," although this definition is still at its infancy and needs more research for further validation. Milk-derived opioid peptides are like morphine and establish naloxone like inhibitory attributes (Gobbetti et al. 2002). Most of these exorphins have a tyrosine amino acid residue at the N-terminal end followed by an aromatic amino acid residue at the 3rd or 4th position (Meisel and FitzGerald 2000). This structural peculiarity enables these exorphins to fit in the binding site of opioid receptors (Teschemacher et al. 1997; Pihlanto-Leppälä 2000; Teschemacher 2003). Casomorphins (α- and β) and lactorphins from milk act as opioid agonists, while casoxins from κ-casein serve as opioid antagonists. Bakalkin and coworkers in 1992 comprehensively described the amino acid composition of exorphins in terms of hydrophilic and hydrophobic amino acid residues. They categorized these residues into four groups. Group P usually includes hydrophobic amino acids including proline, tyrosine, isoleucine, tryptophan, cysteine, and tyrosine. The F group comprises amino acids like phenylalanine, leucine, methionine, and alanine. The hydrophilic amino acids are included in G group (glycine and asparagine) and the B group includes glutamine, glutamic acid, asparagine, and aspartic acid. The group P amino acids are frontrunners instead of followers like glycine or asparagine. The amino acid residues in G group more commonly proceed instead of trailing amino acid residues like phenylalanine, leucine, alanine, and met-group F. The F group amino acids usually precede the group R amino acids (arginine, serine, threonine, and lysine). Finally the R group amino acid residues more commonly precede rather than follow the P group. Overall, the investigators established an association between the primary structure and their physiological response. The main characteristic features are the elements of certain fragment pairs in addition to the density of their arrangement in an opioid peptide (Bakalkin et al. 1992). These observations are supported by the data of Kostyra et al. (2004) with the generated assessment of the hydrophobicity of the bovine-derived exorphins. In general, the literature has shown a total hydrophobicity (2 kcal/mol) and aromatic hydrophobicity (0.6 kcal/mol) for the majority of food exorphins. The total hydrophobicity, as well as aromatic hydrophobicities, established a correlation with the molecular weight of these peptides. The results showed an inverse relationship between the molecular weight and hydrophobicity index. These findings are supported by Bakalkin's observations who demonstrated the presence of hydrophobic domains in food exorphins and endogenous peptides. Moreover, an interesting correlation was established between the presence of aromatic amino acids and the ability to bind the receptors (opioid). The biochemical binding assays, Ki values, and IC$_{50}$ values correlate well with the aromatic hydrophobicity of these peptides (Bakalkin et al. 1992). The above observation regarding the structure of a complex (exorphin-receptor) for effective binding necessitates the tripartite operation of the donor-acceptor organization in addition to a hydrophobic region. The carboxyl and side terminal amino acid residues may help in the development of the donor-acceptor organization. Additionally, the balance between the presence of hydrophobic and hydrophilic amino acid residues may be a supplementary essential aspect (Kostyra et al. 2004).

1.3 Major Food-Derived Exorphins

As discussed in the above section, there is a structural similarity between the endogenous opioids and the food-derived exorphins. Many reports demonstrate human endogenous opioid receptor recognition by food exorphins and regulation of diverse physiological processes.

1.3.1 Casomorphins

Cow milk consists of two major proteins referred to as caseins and whey proteins. The casein is around 80% of bovine milk proteins and there are different subtypes of casein including αs1, αs2, β, and γ. BCM-7 has seven amino acid residues and was first isolated from β-casein of cow milk with morphine-like activity. Later other BCMs (4, 5 and 6) were identified through sequential release of some amino acids from the carboxyl-terminal. More research demonstrated that BCM-5 is the most potent and BCM-7 with prolonged effect among these BCMs (Brantl et al. 1981). BCMs are released from both human and cow milk; however, they differ in the sequence of amino acids. There are very few unique BCMs that are released differentially from either type of milk (Koch et al. 1985). Human milk also produces casomorphin from the α-casein and denoted as hα-casomorphin. Moreover, bovine milk also releases a unique casomorphin peptide known as bovine neocasomorphin-6, with six amino acid residues (Teschemacher et al. 1997; Garg et al. 2016).

1.3.2 Lactorphins and Lactoferroxins

The whey fraction includes proteins like α-lactalbumin, β-lactoglobulin, immunoglobulins, albumin, lactoferrin, etc. The whey has also been a source of exorphins including αb-lactorphin (bovine αb-lactalbumin) and βb-lactorphin (bovine βb-lactoglobulin) (Garg et al. 2016; Teschemacher et al. 1997). The chemically synthesized peptides whose sequences corresponded well with α and β-lactorphins showed angiotensin-I-converting enzyme inhibitory activity (ACE) (Mullally et al. 1997). Nurminen and coworkers have demonstrated that α-lactorphin decreases blood pressure in normotensive and spontaneously hypertensive rats (SHR) (Nurminen et al. 2000). Moreover, other investigators have found that both α and β-lactorphin ameliorated the relaxation of the blood vessels in adult SHR in vitro. The protective outcome of α-lactorphin was focused on endothelial function; however, β-lactorphin showed this response independent of the endothelium (Sipola et al. 2002). Furthermore, reports have established that milk protein-derived α-lactorphin has the potential to up-regulate the expression of the mucin gene (MUC5AC) in the HT29-MTX cell line in vitro. Therefore, it seemingly indicates the role of this exorphin in the development of mucosal immunity in the gut (Martínez-Maqueda et al. 2012).

Lactoferroxins are milk lactoferrin derived opioid peptides. The latter is a globular secretary protein found in various fluids including milk, tears, saliva, and nasal secretions. Three types of exorphins are derived from this protein known as lactoferroxin A, B, and C. All the three forms are carboxymethylated at the carboxyl-terminal. Type A is a lactoferrin fragment from 318 to 325 composed of six amino acid residues with a sequence (YLGSGY-OCH$_3$). It has a molecular formula $C_{32}H_{44}N_6O_{10}$ and molecular weight 672.7 g/mol. Type B is a lactoferrin fragment from 536 to 540 composed of five amino acid residues with a sequence (RYYGY-OCH$_3$). It has a molecular formula $C_{36}H_{46}N_8O_9$ and molecular weight 734.8 g/mol. Type C is a lactoferrin fragment from 673 to 679 composed of seven amino acid residues with a sequence (KYLGPQYOCH$_3$). It has a molecular formula $C_{43}H_{63}N_9O_{11}$ and molecular weight 882 g/mol. The peculiarities of all the three opioid peptides are their antagonistic characteristic feature and have μ-opioid receptor selectivity (Tani et al. 1990).

1.3.3 Casoxins

Casoxins are food exorphins derived from the milk kappa casein. There are four types of casoxins denoted as A, B, C, and D. The first three are derived from kappa casein and the last (D) are derived from alpha (S1) casein. Casoxin A is an opioid peptide having seven amino acids with a sequence of YPSYGLN from the fragment 35 to 41. It has a molecular formula ($C_{47}H_{61}N_9O_{14}$) and molecular weight 976 g/mol. It acts as an opioid antagonist and has opioid receptor selectivity for all the three opioid receptors like MOP, KOP, and DOP. However, the selectivity is far greater towards MOP compared to DOP or KOP. Casoxin B is an opioid peptide having four amino acids with a sequence of YPYY from the fragment 57 to 60. It has a molecular formula ($C_{32}H_{36}N_4O_8$) and molecular weight 604.6 g/mol. It acts as an opioid antagonist and has opioid receptor selectivity for only MOP. Casoxin C is an opioid peptide having four amino acids with a sequence of YIPIQYVLSR from the fragment 25 to 34. It has a molecular formula ($C_{60}H_{94}N_{14}O_{15}$) and molecular weight 1251.5 g/mol. It acts as an opioid antagonist and has opioid receptor selectivity for only MOP. Casoxin D is an opioid peptide having four amino acids with a sequence of YVPFPPF from the fragment 158 to 164. It has a molecular formula ($C_{47}H_{59}N_7O_9$) and molecular weight 866 g/mol. It acts as an opioid antagonist and has opioid receptor selectivity for only MOP or DOP (Chiba et al. 1989; Takahashi et al. 1997; Yoshikawa et al. 1994).

1.3.4 Gluten Exorphins

The exorphins from the wheat are either derived from high molecular weight glutenin or alpha-gliadin. The gluten exorphins (A/B/C) are derived from glutenin and the gliadorphin is derived from the alpha-gliadin. The A-type gluten exorphins both possess common four amino acids (GYYP) from the amino-terminal. The B type gluten exorphins both demonstrate the common first four amino acids (YGGW) from amino-terminal. The gluten exorphin C and gliadorphin both show tyrosine at the N-terminal that is essential for the demonstration of opioid activity. These wheat-derived opioid peptides are derived from the proteins during gastrointestinal digestion in vivo, simulated gastrointestinal digestion (SGID) in vitro or with bacterial fermentation. Reports have established that the patients with neurological manifestations like autism and schizophrenia have often leaky gut that allows these peptides to absorbed more readily, cross the blood–brain barrier (BBB), and then bind opioid receptors in the brain to show various manifestations (Cade et al. 2000). There are various pieces of evidence in favor of this hypothesis and others are against this hypothesis. The formulation of wheat and gluten-free diet is on the basis of this hypothesis (Christison and Ivany 2006).

1.3.5   Rubiscolins

These food-derived opioid peptides also known as exorphins were first discovered in 2001. In an attempt, Yang and coworkers tried to purify these opioid peptides from the pepsin hydrolysates of spinach protein (D-ribulose-1,5-bisphosphate carboxylase/oxygenase (RuBisCo) (Yang et al. 2001). The latter is the most abundant protein of the plant kingdom and is ubiquitous in nature. There are two types of rubiscolins denoted as rubiscolin-5 and rubiscolin-6. The former has a sequence Tyr-Pro-Leu-Asp-Leu and the latter has an amino acid sequence Tyr-Pro-Leu-Asp-Leu-Phe. These opioid peptides have selectivity towards the δ-opioid receptor observed through [^{3}H] deltorphin II-binding assay. These also demonstrate opioid activity observed through mouse vas deferens assay (Yang et al. 2001). The larger opioid peptide (rubiscolin-6) shows a higher affinity for the opioid receptor. The sequence of rubiscolin-6 is found to be conserved and is located in the large subunit of RuBisCo. Reports have shown that when leucine is substituted at position 3rd in the rubiscolin with isoleucine, it increased the δ opioid selectivity (Yang et al. 2003). This substituted opioid peptide does naturally occur in some algae (*Euglena gracilis*, *Euglena stellata*, and *Mesostigma viride*). Therefore, it suggests that the peptide that will be isolated from these algae-based foods may be more potent δ-receptor ligands. Rubiscolin-5 has a molecular formula of $C_{30}H_{45}N_5O_9$ and molecular mass 619.7 g/mol. Rubiscolin-6 has a molecular formula of $C_{39}H_{54}N_6O_{10}$ and molecular mass 766.9 g/mol.

1.3.6   Soymorphins

The soymorphins are usually chemically synthesized peptides that have a sequence similar to the peptide Tyr-Pro-Phe-Val that is derived from the β-conglycinin β-subunit of soy. The sequence analysis shows that this sequence is similar to human βh-casomorphin-4. Soymorphin-5 has a sequence Tyr-Pro-Phe-Val-Val, soymorphin-6 Tyr-Pro-Phe-Val-Val-Asn, and soymorphin-7 Tyr-Pro-Phe-Val-Val-

Asn-Ala. Soymorphins have selectivity towards the μ-opioid receptor. Among these three soymorphins, the soymorphins-5 demonstrates the highest opioid activity (Ohinata et al. 2007). Soymorphin-5 has molecular mass 623.3308 g/mol, soymorphin-6 has 737.3736 g/mol, and soymorphin-7 with a molecular mass of 808.4106 g/mol. More research is needed to analyze the production of these peptides in large quantities and their opioid activity.

1.3.7   Oryzatensin

Researchers have isolated a novel peptide the hydrolysate of soluble protein of rice (albumin). The hydrolysis was performed with trypsin and has a sequence Gly-Tyr-Pro-Met-Tyr-Pro-Leu-Pro-Arg. It was demonstrated to have anti-opioid activity when observed through GPI assay. This peptide depicted biphasic contractions on ileum that showed rapid followed by slower contraction. When inhibitors like atropine and tetrodotoxin were used in the assay, it was observed that slower contractions were mediated by the cholinergic nervous system. The results showed that this peptide has a weak affinity towards μ-receptors and linked with slower contraction. This peptide was also found to demonstrate phagocytosis-promoting activity in polymorphonuclear leukocytes derived from humans. This rice derived exorphin was found to amplify the generation of superoxide anion in vitro in peripheral leukocytes (human) (Takahashi et al. 1994).

1.4   Receptors for Food-Derived Exorphins

The opioid receptors are transmembrane proteins belonging to a large family of G protein-coupled receptors (GPCRs). GPCR signaling occurs through the involvement of some membrane-bound enzymes like adenylyl cyclase, phospholipase C, and membrane channels (calcium) (Waldhoer et al. 2004). These receptors are located on the gastrointestinal tract, immune cells, endocrine cells, peripheral and central nervous systems (Wittert et al. 1996). Opioid receptors are classified into three main types including μ (MOP), κ (KOP), and δ (DOP). These receptors have almost 65% identity with each other and span the membrane seven times to form three extracellular and three intracellular loops. The helical segments (seven) demonstrate common characteristics and result in the formation of an opioid-binding region. Moreover, it results in the formation of divergent features that define great affinity and specificity for other deeper sub-type ligands (Quock et al. 1999). When these exogenous food-derived peptides bind with these types of receptors on different cells, these result in various biological responses including a decrease of gastrointestinal transit, anxiolytic activities, hormone secretion, analgesic effects, depression in respiration, euphoria, role in fat intake, effects on metabolism, and others. More generally, agonists that are specific for μ and δ-opioid receptors show analgesic effects and therefore gratifying pain; however, κ-opioid receptor agonists are dysphoric (Waldhoer et al. 2004). The opioid peptides derived from various foods demonstrate affinities towards different opioid receptors as shown in Table 1.1.

1.4.1   μ-Opioid Receptor (MOP)

More than 40 years ago the investigators established that food protein after gastrointestinal digestion releases peptides that have theoretically a potential for binding with various endogenous opioid receptors (Zioudrou et al. 1979). For the first time, these types of atypical food-derived exorphins were reported from milk in 1970 and were named as BCMs (Brantl et al. 1979, 1981, 1982; Zioudrou et al. 1979; Koch et al. 1985). These exorphins were demonstrated to have an affinity for the μ-type opioid receptor, although some members showed an affinity for δ-opioid receptors also. Moreover, human BCMs were found to be less potent compared to bovine BCMs (Koch et al. 1985). BCM-5 showed the highest and BCM-7 the least activity leaving in the middle BCM-4 and BCM-8. Bovine BCMs (4-7) showed agonistic activity for the μ-type opioid receptors; however, the bovine BCM-8 demonstrated specificity for all the three receptors. The human BCMs (4, 5, and 7) showed agonistic activity towards all three types of receptors. On the contrary, the casoxins A, B, C, and D demonstrated receptor antagonistic activity, B and C showed activity towards μ-type opioid receptors while casoxin A depicted activity towards all the three types of receptors and casoxin D showed affinity towards μ and δ-opioid receptors. The bovine α- and β-lactorphins demonstrated their agonistic activities towards μ-type opioid receptors. The opioid peptides derived from milk lactoferrin as lactoferroxin demonstrate antagonistic activity for μ-type opioid receptors. The gluten exorphins A, B, and C act as agonists. Type A gluten exorphin (A4, A5) and C demonstrate μ and δ and type B (B4, B5) μ-opioid receptor activation. The soymorphins-5, 6, and 7 released from the soy act as opioid receptor agonists and demonstrate activity for μ-type opioid receptors.

1.4.2   δ-Opioid Receptor (DOP)

These types of opioid receptors were one of the first GPCR that demonstrated constitutive activity (Costa and Herz 1989). Through the involvement of agonist binding responses, some of the antagonist affinities may be controlled significantly through the participation of ions and analogs of energy-rich species like GTP/GDP (Neilan et al. 1999). Although it has been reported that there are not any direct regulators (allosteric) that can control δ-opioid receptors, however, studies have established that GPCR kinases, G proteins, and β-arrestins evidently control the affinities and functions of these receptors. Moreover, guanyl nucleotides and sodium have an ability to modify the interaction between G protein and δ-opioid receptor complex and interaction. Additionally, another GPCR has the potential to result in the formation of heterodimers with δ-opioid receptors possibly transforming δ-opioid activity (Rios et al. 2004). The food exorphins released from the spinach

Table 1.1 Food-derived opioid peptides and their receptors (Garg et al. 2016)

Food	Proteins		Opioid Receptor Selectivity	Antagonist/ antagonist	References
Bovine milk	β-casein	Bovine BCM-4	μ	A	Brantl et al. (1981)
		Bovine BCM-5	μ	A	
		Bovine BCM-6	μ	A	
		Bovine BCM-7	μ	A	
		Bovine BCM-8	μ/κ/δ	A	Teschemacher et al. (1997)
	κ-casein	Casoxin A	μ/κ/δ	AN	Chiba et al. (1989)
		Casoxin B	μ	AN	
		Casoxin C	μ	AN	
		Casoxin D	μ/δ	AN	Yoshikawa et al. (1994)
	α-lactalbumin	Bovine α-lactorphin	μ	A	Teschemacher et al. (1997)
	β-lactoglobulin	Bovine β-lactorphin	μ	A	Day et al. (1981)
	Lactoferrin	Lactoferroxin A	μ	AN	Tani et al. (1990)
		Lactoferroxin B	μ	AN	
		Lactoferroxin C	μ	AN	
Wheat	Glutenin	Gluten exorphin A5	μ/δ	A	Yoshikawa et al. (2003) Fukudome and Yoshikawa (1992)
		Gluten exorphin A4	μ/δ	A	
		Gluten exorphin B5	δ	A	
		Gluten exorphin B4	δ	A	
		Gluten exorphin C	μ/δ	A	Yoshikawa et al. (2003) Fukudome and Yoshikawa (1992)
Spinach	RuBisCo	Rubiscolin-5	δ	A	Yang et al. (2001)
		Rubiscolin-6	δ	A	
Soy	β-conglycinin	Soymorphin-5	μ	A	Ohinata et al. (2007)
		Soymorphin-6	μ	A	

(continued)

Table 1.1   (continued)

Food	Proteins		Opioid Receptor Selectivity	Antagonist/ antagonist	References
		Soymorphin-7	μ	A	
Rice	Soluble protein (Albumin)	Oryzatensin	μ	AN	Takahashi et al. (1994) Yoshikawa (2015) Jinsmaa et al. (2001)

RuBisCo rubiscolin-5 and rubiscolin-6 have been found to show biological effects through δ-opioid receptors. The exorphins released from the wheat of type B (B4 and B5) also show their affinity for these types of receptors. However, type A (A4, A5) gluten exorphins have more affinity towards δ in comparison to the μ-opioid receptors. Similarly, gluten exorphin C has been demonstrated to have more affinity towards δ compared to μ-opioid receptors. The casoxin D generated from the bovine milk κ-casein has also shown an affinity for both the μ and δ-opioid receptors; however, casoxin A has an affinity for μ, κ, and δ-opioid receptors. The human BCMs (4, 5, 7) have an affinity for all three types of opioid receptors.

1.4.3   κ-Opioid Receptor (KOP)

These types of receptors are fairly distributed and represent only a fraction of the total receptors present in the brain. In contrast to the other species, the guinea pig brain possesses more κ-opioid receptors compared to the other two types. These receptors are further divided into subtypes of κ_1 and κ_2. Studies have shown that peptides and benzeneacetamides are κ_1 agonists and benzomorphans have been found to bind with both κ_1 and κ_2-opioid receptors. The primary transduction mechanism involves G_i/G_o family as the transducer and the effector or response involves adenylate cyclase inhibition of potassium and calcium channels (Attali et al. 1989). The secondary transduction mechanism involves G_i/G_o and G_{12}/G_{13} family as the transducer and the effector or response involves adenylate cyclase inhibition and phospholipase C stimulation (Bohn et al. 2000). Human BCMs (4, 5, 7, 8) and bovine BCM-8 demonstrate their affinity for κ-receptors, although this affinity is weaker compared to μ and δ-opioid receptors. Moreover, the human α-BCM exhibits far greater affinity towards κ-receptors compared to the other two opioid receptors. The casoxin A derived from κ-casein also shows an affinity towards this opioid receptor.

1.5   Production of Food-derived Exorphins

The proteins in various foods like milk, wheat, rice, and soy do not possess opioid activity. This activity is latent in the encrypted BAPs in the intact proteins. Once these proteins are digested through various gastrointestinal enzymes or fermentation, they release peptides of around 4–20 amino acids that have an affinity for various opioid receptors and demonstrate numerous biological effects (Korhonen and Pihlanto 2006; Udenigwe et al. 2012; Choi et al. 2012). To detect these peptides in the intestinal fluid it is very cumbersome, therefore simulated gastrointestinal digestion with pepsin, trypsin, chymotrypsin, elastase, leucine aminopeptidase, and carboxypeptidase is performed at corresponding pH and physiological temperature. Overall, SGID and fermentation are prerequisites for the analysis of the production of exorphins from various dietary biomolecules.

1.5.1   Fermentation

Many daily dietary or nutritional products are manufactured through the fermentation process including sauerkraut, yogurt, salami, kimchi, bread, and beverages. Quite evidently, the fermentation process exploits the use of microorganisms. These microbes secrete in the medium the proteolytic enzymes that degrade various biomolecules. Among the various biomolecules, the proteins are hydrolyzed into individual amino acids and some intact peptides. Many of these peptides were demonstrated to have opioid activity. Immunoreactive materials resembling that of BCM were recovered from fermented milk (*Bacillus cereus* or *Pseudomonas aeruginosa*) (Hamel et al. 1985). Pre-treatment of the fermented milk (*Lactobacillus GG*) with the proteolytic system (pepsin and trypsin) also resulted in the production of various exorphins (Rokka et al. 1997). Belem and coworkers in 1999 fermented whey with the *Kluyveromyces marxianus var. marxianus* and detected the release of YLLF that corresponds to the milk exorphin β-lactorphin released from digestion of β-lactoglobulin (Belem et al. 1999). As described in the previous section that prolyl peptidases like DPPIV hydrolyzing food-derived exorphins because of the presence of proline residues for which these endopeptidases have specificity. In this perespective, the bacterial strains that are deficient for the genes that code these enzymes (PepX) when used for the fermentation of milk, resulting in BCM (4) production (Matar and Goulet 1996). Although Schieber and coworkers could not detect any intact BCM peptide, nevertheless, BCM precursors were detected when yogurt was fermented with *S. salivarius ssp. thermophilus* and *L. delbrueckii ssp. bulgaricus* (Schieber and Brückner 2000). The possible explanation may be the release of oligo prolyl proteases by the bacteria that may have digested the BCM sequences. Alternately, it may be the ability of the stature cultures to cleave the β-casein and produce the BCM sequences. Additionally, the more probable reason may the production of BCM sequences by *L. delbrueckii ssp. Bulgaricus* and simultaneous degradation *by S. thermophilus*. It is also possible that the milk used in this fermentation is A2 "like" that is not the source of BCMs. Regarding the cheese, BCM-7 was not detected in the Brie and Cheddar and the possible reason may be the use of A1 "like" milk or the sensitivity of the analytical techniques employed for the detection of these opioid peptides. It is also probable that BCMs may be degraded during ripening as advocated by the fact that enzymes secreted by *L. lactis* ssp. *cremoris* decrease the concentration of this opioid peptide (BCM-7) by half (1.5% NaCl and pH 5.0) after 42–105 days (Muehlenkamp and Warthesen 1996). Jarmołowska and coworkers for the first time detected BCM-7 in Brie cheese and the latter other investigators recovered 6.48 to 0.15 µg/g BCM-7 (Sienkiewicz-Szłapka et al. 2009; De Noni and Cattaneo 2010). BCM-7 was also assessed in a number of cheese samples including Cheddar cheeses, Gorgonzola, Gouda and Fontina and the amount of BCM-7 ranges from 0.01 - 0.11 µg/g. The SGID treatment further increased the yield up to 21.77 and 15.22 µg/g Gouda and Cheddar cheeses, respectively (De Noni and Cattaneo 2010). Moreover, opioid agonists (BCM) and antagonists (casoxin 6, casoxin C, and lactoferroxin A) were identified in Gouda, Adamski, and Kasztelan cheeses (Sienkiewicz-Szłapka et al. 2009). These all investigations highlight the formation of food exorphins fermented by bacteria and fungus because of the secretion of various combinations of enzymes by these microorganisms.

1.5.2   Simulated Gastrointestinal Digestion (SGID)

In SGID the proteins from various food sources serve as the substrates like casein, lactoferrin, α-lactalbumin, and β-lactoglobulin of milk, gluten of wheat, RuBisCo of spinach, β-conglycinin of soybean and albumin of rice. The combination of enzymes (pepsin, trypsin, chymotrypsin, elastase, leucine aminopeptidase) is exploited for the hydrolysis protocols. The pH and temperature are simulated to that of the human temperature. We have also exploited this protocol in our laboratory for the hydrolysis of cow milk protein. The β-casein variants of cow milk (A1A1, A1A2, and A2A2) were isolated from total casein by urea fractionation from bovine crossbred Karan Fries animals. The alternate variants of β-casein were subjected to SGID in vitro. The analysis of BCMs was carried out in two fractions collected on reverse phase preparative HPLC. The MS-MS data showed the precursor of BCM in the form of 14 amino acid peptide. When the same precursor containing fraction was further digested with elastase and leucine aminopeptidase, BCM-7 was detected by competitive ELISA. BCM-7 could not be detected from the homozygous A2A2 milk variants. More importantly, BCM-5 was not recovered from any of the β-casein variants. The amount of BCM-7 recovered from A1A1 was 3.2 times more than the A1A2 heterozygous β-casein of cow milk. The opioid activity of the released peptides was detected on the rat ileum on an isolated organ bath. The opioid activity was expressed in terms of IC_{50} values and the estimated values were 0.534–0.595 µM for A1A1 and 0.410–0.420 µM for A1A2 (Raies et al. 2015). The generations of BCMs after SGID by various enzymes have been analyzed in milk and other dairy products (De Noni and Cattaneo 2010) and differential release has further been detected in various variants of cow β-casein (Cieslinska et al. 2007;

De Noni 2008; Raies et al. 2015). The exploitation of SGID with proteolytic enzymes and microbial enzymes thermolysin and alcalase was performed for the assessment of the generation of food exorphins. Pepsin digestion alone resulted in the production of food exorphins and the addition of other enzymes increased the yield of these exorphins (Zioudrou et al. 1979). Fukudome and Yoshikawa in 1992 detected the release of A4, A5, B4, and B5 in hydrolysates digested with pepsin-thermolysin (Fukudome and Yoshikawa 1992). On a comparative basis, the release of gluten derived opioid peptide exorphin A5 was higher in hydrolysate with pepsin-elastase (250 µg/g) compared to pepsin-thermolysin (40 µg/g) (Fukudome et al. 1997). The gluten exorphins A5 and C were produced from bread gluten (0.747–2.192 mg/kg) and pasta gluten (3.201–6.689 mg/kg), respectively, after SGID with pepsin-trypsin-chymotrypsin (Stuknytė et al. 2015). In 2008, Kong and coworkers detected the opioid activity of hydrolysates with alcalase for 6 h and pepsin for 24 h. The IC_{50} values were 1.21 and 1.57 mg/ml for alcalase and pepsin hydrolysates, respectively (Kong et al. 2008).

1.6   Absorption and Fate of Exorphins

The food proteins are digested in the gastrointestinal tract by a combination of various proteolytic enzymes. These biocatalysts may release amino acids serially from ends (exopeptidases) or hydrolyze the peptide bonds within the protein (non-terminal amino acids). Carboxypeptidases are examples of exopeptidases and pepsin and trypsin are examples of endopeptidases. The combinations of these proteases like pepsin, trypsin, chymotrypsin, elastase, and carboxypeptidase and leucine aminopeptidases cleave the proteins to their individual amino acids and absorbed through the intestine. There are some proline-rich peptides that escape this hydrolysis, but some brush border enzymes like prolyl oligopeptidase including dipeptidyl-peptidase IV (DDPIV) is specific for these proline peptides. Some of the peptides with an amino acid composition of 4–20 escape this hydrolysis and find a chance to get absorbed in the intestine, reach the blood, and may cross the blood–brain barrier (BBB). These peptides may have physiological roles in the gastrointestinal tract, blood cells, immune cells, and central nervous system (CNS). The food exorphins as discussed in the above section usually are proline-rich fragments produced during the gastrointestinal tract and find a chance to reach the blood and brain. However, the absorption of these exorphins in the gut depends upon many factors including the size of these peptides (smaller the peptide more chances of absorption), pKa of these peptides, and finally the pH of the microenvironment. The processes like intestinal transit and gastric emptying have an impact on the absorption of these peptides. Studies have established that in normal individuals peptides fragments exceeding two or three amino acids find it difficult to get absorbed in the intestine. Physiological complications like gut inflammation, leaky gut conditions, and stress lead to increased gut permeability and in turn an increased absorption of these peptides in the intestine (EFSA 2009). These peptides are absorbed in the intestinal epithelium through paracellular or transcellular pathways. Many reports have already established that transcellular absorption is mediated through the carrier, endocytosis or transcytosis. Further literature has demonstrated that intestinal mucosa is unable to provide a specific barrier and therefore numerous peptide fragments like BCMs, gliadorphins, lactorphins, and soymorphins are absorbed through intestinal epithelium (Stuknytė et al. 2015; Chabance et al. 1998; Roumi et al. 2001; Quirós et al. 2008; Shimizu et al. 1997; Dorkoosh et al. 2004). Although there is not sufficient data available that clearly demonstrates the transport of opioid peptides across the gastrointestinal tract. Nevertheless, the transport of a synthetic opioid peptide DADLE has been observed to occur through sodium-coupled intestinal transporter (Ananth et al. 2012). Moreover, other researchers have found that opioid peptides are unable to transport through the cellular membranes, paracellular pathways, or diffusion (Patel et al. 2007; Iwan et al. 2008). The absorption of opioid peptides across the intestine has been demonstrated through in vitro system in cell lines and in vivo in various animal models. Immunoreactive material resembling BCM-7 has been detected in the plasma of calves (newborn) after milk consumption (Umbach et al. 1985). The opioid peptide morphiceptin has been found to activate the opioid receptor only when the addition was performed from the serosal side of rabbit ileum in comparison to the mucosal side. It was concluded that brush border enzymes like prolyl oligopeptidase including DPPIV cleave this proline-rich opioid peptide. When this peptide was co-administered with the DPPV inhibitor diisopropylfluorophosphate (DIPFP), the degradation of this peptide with this enzyme was inhibited (Mahe et al. 1989a, b). The other investigators have established the trans-epithelial transport of BCM-7 and BCM-5 that act as agonists of the μ-opioid receptor. As already mentioned in the above section, these peptides are released from bovine β-casein of milk protein. The same investigators depicted the transport of μ-opioid receptor antagonist lactoferroxin A in the same model, the latter is produced from lactoferrin of milk protein (Iwan et al. 2008).

1.7 Transport to Blood and Across Blood–Brain Barrier (BBB)

The half-life of the food exorphins is very short in the blood, due to the presence of prolyl oligopeptidase like DPPIV. The opioid peptides of endogenous nature like leuenkephalin and dynorphin (1–13) show half-life of less than 6 and 1 min, respectively (Brugos and Hochhaus 2004; Wang et al. 2006). However, Negri and coworkers have found a longer half-life of Dermorphin in comparison to the enkephalins (Negri et al. 1998). The other investigators have suggested that the half-life can be enhanced if these peptides are able to bind the carrier proteins like iron-containing transferrin and albumin (Dennis et al. 2002; Kim et al. 2010). The half-life of the exorphins in the blood has not still been explored deeper and deserves further investigation. Ganapathy and coworkers in 2005 have found that the exorphins can be transported from blood to the central nervous system (CNS) through BBB by transport systems from PTS-1 to PTS-4 (Ganapathy and Miyauchi 2005). Peptides like leuenkephalin and Tyr-MIF-1 met-enkephalin are transported via PTS-1 while some elements are expressed at the BBB like organic anion transporting polypeptides that are used by deltorphin II and DPDPE for their transport (Gao et al. 2000). Similarly, Brasnjevic and coworkers in 2009 found that amino acid carrier (neutral) is used for the transport of potent opioid analgesic known as biphasic (Brasnjevic et al. 2009). Moreover, the glycosylated peptides were found to be transported through GLUT1 (glucose transporter) as well as glycosylated analogs of dermorphin and met-encephalin (Brasnjevic et al. 2009).

References

Ananth S, Thakkar SV, Gnana-prakasam JP et al (2012) Transport of the synthetic opioid peptide DADLE ([d-Ala2, d-Leu5]–Enkephalin) in neuronal cells. J Pharm Sci 101(1):154–163

Attali B, Saya D, Vogel Z (1989) Kappa-opiate agonists inhibit adenylate cyclase and produce heterologous desensitization in rat spinal cord. J Neurochem 52:360–369

Bakalkin G, Ya Demuth HU, Nymberg F (1992) Relationship between primary structure and activity in exorphins and endogenous opioid peptides. FEBS Lett 310:13–16

Belem MA, Gibbs BF, Lee BH et al (1999) Proposing sequences for peptides derived from whey fermentation with potential bioactive sites. J Dairy Sci 82(3):486–493

Bhat ZF, Kumar S, Bhat HF (2015) Bioactive peptides from egg: a review. Nutr Food Sci 45:190–212

Birgisdottir BE, Hill JP, Harris DP et al (2002) Variation in consumption of cow milk proteins and lower incidence of type 1 diabetes in Iceland vs the other 4 Nordic countries. Diabetes Nutr Metab Dis 15(4):240–245

Bohn LM, Belcheva MM, Coscia CJ (2000) Mitogenic signaling via endogenous kappa-opioid receptors in C6 glioma cells: evidence for the involvement of protein kinase C and the mitogen-activated protein kinase signaling cascade. J Neurochem 74:564–573

Bojović K, Stanković B, Kotur N (2017) Genetic predictors of celiac disease, lactose intolerance, and vitamin D function and presence of peptide morphins in urine of children with neurodevelopmental disorders. Nutr Neurosci 22(1):40–50

Brandt W, Barth A, Höltje HD (1994) Investigations of structure-activity relationships of β-casomorphins and further opioids using methods of molecular graphics. In: Brantl V, Teschemacher H (eds) β-casomorphins and related peptides: recent developments. Wiley, Weinheim, pp 93–104

Brantl V, Teschemacher H, Henschen A et al (1979) Novel opioid peptides derived from casein (beta-casomorphins). I. Isolation from bovine casein peptone. Hoppe Seylers Z Physiol Chem 360(9):1211–1216

Brantl V, Teschemacher H, Blasig J et al (1981) Opioid activities of beta-casomorphins. Life Sci 28 (17):1903–1909

Brantl V, Pfeiffer A, Herz A et al (1982) Antinociceptive potencies of beta-casomorphin analogs as compared to their affinities towards mu and delta opiate receptor sites in brain and periphery. Peptides 3(5):793–797

Brasnjevic I, Steinbusch HW, Schmitz C et al (2009) Delivery of peptide and protein drugs over the blood–brain barrier. Prog Neurobiol 87(4):212–251

Brugos B, Hochhaus G (2004) Metabolism of dynorphin A (1-13). Pharmazie 59(5):339–343

Cade R, Privette R, Fregly M et al (2000) Autism and schizophrenia: intestinal disorders. Nutr Neurosci 3:57–72

Carrasco-Castilla J, Hernández-Álvarez AJ, Jiménez-Martínez C et al (2012) Use of proteomics and peptidomics methods in food bioactive peptide science and engineering. Food Eng Rev 4:224–243

Chabance B, Marteau P, Rambaud JC et al (1998) Casein peptide release and passage to the blood in humans during digestion of milk or yogurt. Biochimie 80(2):155–165

Chiba H, Tani F, Yoshikawa M (1989) Opioid antagonist peptides derived from κ-casein. J Dairy Res 56:363–366

Choi J, Sabikhi L, Hassan A, Anand S (2012) Bioactive peptides in dairy products. Int J Dairy Technol 65(1):1–12

Christison GW, Ivany K (2006) Elimination diets in autism spectrum disorders: any wheat amidst the chaff? J Dev Behav Pediatr 27(2):S162

Cieslinska A, Kaminski S, Kostyra E et al (2007) β-Casomorphin-7 in raw and hydrolyzed milk derived from cows of alternative β-casein genotypes. Milchwissenschaft 62:125–127

Costa T, Herz A (1989) Antagonists with negative intrinsic activity at delta opioid receptors coupled to GTP-binding proteins. Proc Natl Acad Sci 86(19):7321–7325

Day A, Freer R, Liao C (1981) Morphiceptin (beta-casomorphin (1-4) amide): a peptide opioid antagonist in the field stimulated rat vas deferens. Res Commun Chem Pharmacol 34 (3):543–546

De Noni I (2008) Release of beta-casomorphins 5 and 7 during simulated gastro-intestinal digestion of bovine beta-casein variants and milk-based infant formulas. Food Chem 110(4):897–903

De Noni I, Cattaneo S (2010) Occurrence of b-casomorphins 5 and 7 in commercial dairy products and in their digests following in vitro simulated gastro-intestinal digestion. Food Chem 119:560–566

Dennis MS, Zhang M, Meng YG et al (2002) Albumin binding as a general strategy for improving the pharmacokinetics of proteins. J Biol Chem 277(38):35035–35043

Dorkoosh FA, Broekhuizen CAN, Borchard G et al (2004) Transport of octreotide and evaluation of mechanism of opening the paracellular tight junctions using superporous hydrogel polymers in Caco-2 cell monolayers. J Pharm Sci 93(3):743–752

European Food Safety Authority (EFSA) (2009) Review of the potential health impact of β-casomorphins and related peptides. Sci Rep 231:1–107

Elliott RB, Harris DP, Hill JP et al (1999) Type 1 (insulin-dependent) diabetes mellitus and cow milk: casein variant consumption. Diabetologia 42:292–296

Fields K, Falla TJ, Rodan K et al (2009) Bioactive peptides: signaling the future. J Cosmet Dermatol 8:8–13

Fukudome S, Yoshikawa M (1992) Opioid peptides derived from wheat gluten: their isolation and characterization. FEBS Lett 296(1):107–111

Fukudome S, Jinsmaa Y, Matsukawa T (1997) Release of opioid peptides, gluten exorphins by the action of pancreatic elastase. FEBS Lett 412(3):475–479

Ganapathy V, Miyauchi S (2005) Transport systems for opioid peptides in mammalian tissues. AAPS J 7(4):E852–E856

Gao B, Hagenbuch B, Kullak-Ublick GA et al (2000) Organic anion-transporting polypeptides mediate transport of opioid peptides across blood-brain barrier. J Pharmacol Exp Ther 294 (1):73–79

Garg S, Nurgali K, Mishra V (2016) Food proteins as source of opioid peptides-A review. Curr Med Chem 23(9):893–910

Gobbetti M, Stepaniak L, De Angelis M et al (2002) Latent bioactive peptides in milk proteins: proteolytic activation and significance in dairy processing. Crit Rev Food Sci Nutr 42 (3):223–239

Hamel U, Kielwein G, Teschemacher H (1985) Beta-casomorphin immunoreactive materials in cows' milk incubated with various bacterial species. J Dairy Res 52(1):139–148

Haque E, Chand R, Kapila S (2008) Biofunctional properties of bioactive peptides of milk origin. Food Rev Int 25:28–43

Iwan M, Jarmołowska B, Bielikowicz K et al (2008) Transport of μ-opioid receptor agonists and antagonist peptides across Caco-2 monolayer. Peptides 29(6):1042–1047

Jinsmaa Y, Takenaka Y, Yoshikawa M (2001) Designing of an orally active complement C3a agonist peptide with anti-analgesic and anti-amnesic activity. Peptides 22(1):25–32

Kim BJ, Zhou J, Martin B et al (2010) Transferrin fusion technology: a novel approach to prolonging biological half-life of insulinotropic peptides. J Pharmacol Exp Ther 334 (3):682–692

Kitts DD, Weiler K (2003) Bioactive proteins and peptides from food sources. Applications of bioprocesses used in isolation and recovery. Curr Pharm Des 9:1309–1323

Koch G, Wiedemann K, Teschemacher H (1985) Opioid activities of human β-casomorphins. N-S Arch Pharmacol 331(4):351–354

Kong X, Zhou H, Hua Y et al (2008) Preparation of wheat gluten hydrolysates with high opioid activity. Eur Food Res Technol 227(2):511–517

Korhonen H (2009) Milk-derived bioactive peptides: from science to applications. J Funct Foods 1:177–187

Korhonen H, Pihlanto A (2003) Food-derived bioactive peptides-Opportunities for designing future foods. Curr Pharm Des 9:1297–1308

Korhonen H, Pihlanto A (2006) Bioactive peptides: production and functionality. Int Dairy J 16 (9):945–960

Kostyra E, Sienkiewicz-Szapka E, Jarmolowska B et al (2004) Opioid peptides derived from milk proteins. Pol J Food Nutr 13(54):25–35

Laugesen M, Elliott R (2003a) Ischaemic heart disease, type 1 diabetes, and cow milk A1 β-casein. N Z Med J 116(1168):U295

Laugesen M, Elliott R (2003b) The influence of consumption of A1 β-casein on heart disease and type 1 diabetes–the authors reply. Z Med J 116(1170):U367

Lázaro CP, Pondé MP, Rodrigues LEA (2016) Opioid peptides and gastrointestinal symptoms in autism spectrum disorders. Rev Bras Psiquiatr 38(3):243–246

Lemes AC, Sala L, Ores JDC et al (2016) A review of the latest advances in encrypted bioactive peptides from protein-rich waste. Int J Mol Sci 17(6):950

Liebmann C, Barth A, Neubert K et al (1986) Effects of β-Casomorphin on 3h-ouabain binding to guinea-pig heart membranes. Pharamzie 41:670–671

Mahe S, Tome D, Dumontier AM et al (1989a) Absorption of intact morphiceptin by diisopropylfluorophosphate-treated rabbit ileum. Peptides 10(1):45–52

Mahe S, Tome D, Dumontier AM et al (1989b) Absorption of intact beta-casomorphins (beta-CM) in rabbit ileum in vitro. Reprod Nutr Dev 29(6):725–733

Martínez-Maqueda D, Miralles B, De Pascual-Teresa S et al (2012) Food-derived peptides stimulate mucin secretion and gene expression in intestinal cells. J Agric Food Chem 60(35):8600–8605

Matar C, Goulet J (1996) β-casomorphin 4 from milk fermented by a mutant of Lactobacillus helveticus. Int Dairy J 6(4):383–397

McLachlan CNS (2001) β-Casein A1, ischaemic heart disease mortality, and other illnesses. Med Hypothesis 56:262–272

Meisel H, FitzGerald RJ (2000) Opioid peptides encrypted in intact milk protein sequences. Br J Nutr 84(1):27–31

Moldes AB, Vecino X, Cruz JM (2017) Nutraceuticals and food additives. In: Pandey A, Du G, Sanroman MA et al (eds) Current developments in biotechnology and bioengineering: food and beverages industry. Elsevier, Amsterdam, pp 143–164

Moller NP, Scholz-Ahrens KE, Roos N et al (2008) Bioactive peptides and proteins from foods: indication for health effects. Eur J Nutr 47(4):171–182

Monetini L, Cavallo MG, Manfrini S (2002) Antibodies to bovine β-casein in diabetes and other autoimmune diseases. Horm Metab Res 34(8):455–459

Mora L, Reig M, Toldra F (2014) Bioactive peptides generated from meat industry by-products. Food Res Int 65:344–349

Muehlenkamp MR, Warthesen JJ (1996) Beta-casomorphins: analysis in cheese and susceptibility to proteolytic enzymes from Lactococcus lactis ssp cremoris. J Dairy Sci 79(1):20–26

Mullally MM, Meisel H, FitzGerald RJ (1997) Identification of a novel angiotensin-I-converting enzyme inhibitory peptide corresponding to a tryptic fragment of bovine b-lactoglobulin. FEBS Lett 402:99–101

Negri L, Lattanzi R, Tabacco F et al (1998) opioid peptides with potent and prolonged analgesic activity and enhanced blood-brain barrier penetration. Br J Pharmacol 124(7):1516–1522

Neilan CL, Akil H, Woods JH et al (1999) Constitutive activity of the delta-opioid receptor expressed in C6 glioma cells: identification of non-peptide delta-inverse agonists. Br J Pharmacol 128:556–562

Nurminen ML, Sipola M, Kaarto H et al (2000) α-Lactorphin lowers blood pressure via radiotelemetry in normotensive and spontaneously hypertensive rats. Life Sci 66:1535–1543

Ohinata K, Agui S, Yoshikawa M (2007) Soymorphins, novel μ opioid peptides derived from soy β-conglycinin β-subunit, have anxiolytic activities. Biosci Biotechnol Biochem 71 (10):2618–2621

Patel LN, Zaro JL, Shen WC (2007) Cell penetrating peptides: intracellular pathways and pharmaceutical perspectives. Pharm Res 24(11):1977–1992

Pihlanto-Leppälä A (2000) Bioactive peptides derived from bovine whey proteins. Trends Food Sci Technol 11:347–356

Przybylski R, Firdaous L, Châtaigné G et al (2016) Production of an antimicrobial peptide derived from slaughterhouse byproduct and its potential application on meat as preservative. Food Chem 211:306–313

Quirós A, Dávalos A, Lasunción MA et al (2008) Bioavailability of the antihypertensive peptide LHLPLP: transepithelial flux of HLPLP. Int Dairy J 18(3):279–286

Quock RM, Burkey TH, Varga E et al (1999) The d-opioid receptor: molecular pharmacology, signal transduction, and the determination of drug efficacy. Pharmacol Rev 51(3):503–532

Raies MH, Kapila R, Sharma R et al (2014a) Comparative evaluation of cow β-casein variants (a1/a2) consumption on th2-mediated inflammatory response in mouse gut. Eur J Nutr 53 (4):1039–1049

Raies MH, Kapila R, Saliganti V (2014b) Consumption of β-casomorphins-7/5 induce inflammatory immune response in mice gut through th2 pathway. J Funct Foods 8:150–160

Raies MH, Kapila R, Kapila S (2015) Release of β-casomorphin-7/5 during simulated gastrointestinal digestion of milk β-casein variants from Indian crossbred cattle (Karan fries). Food Chem 1 (168):70–79

Rios C, Gomes I, Devi LA (2004) Interactions between delta opioid receptors and alpha-adrenoceptors. Clin Exp Pharmacol Physiol 31:833–836

Rokka T, Eeva-Liisa S, Jari T et al (1997) Release of bioactive peptides by enzymatic proteolysis of Lactobacillus GG fermented UHT milk. Milchwissenschaft 52:675–678

Roumi M, Kwong E, Deghenghi R et al (2001) Permeability of the peptidic GH secretagogues hexarelin and EP 51389, across rat jejunum. Peptides 22(7):1129–1138

Sánchez A, Vázquez A (2017) Bioactive peptides: a review. Food Qual Saf 1:29–46

Schieber A, Brückner H (2000) Characterization of oligo-and polypeptides isolated from yoghurt. Eur Food Res Technol 210(5):310–313

Shahidi F, Zhong Y (2008) Bioactive peptides. J AOAC Int 91(4):914–931

Sharma S, Singh R, Rana S (2011) Bioactive peptides: a review. Int J Bioautom 15:223–250

Shimizu M, Tsunogai M, Arai S (1997) Transepithelial transport of oligopeptides in the human intestinal cell, CACO-2. Peptides 18(5):681–687

Sienkiewicz-Szłapka E, Jarmołowska B, Krawczuk S et al (2009) Contents of agonistic and antagonistic opioid peptides in different cheese varieties. Int Dairy J 19(4):258–263

Singh BP, Vij S, Hati S (2014) Functional significance of bioactive peptides derived from soybean. Peptides 54:171–179

Sipola M, Finckenberg P, Korpela R et al (2002) Effect of long-term intake of milk products on blood pressure in hypertensive rats. J Dairy Res 69:103–111

Stuknytė M, Maggioni M, Cattaneo S et al (2015) Release of wheat gluten exorphins A5 and C5 during in vitro gastrointestinal digestion of bread and pasta and their absorption through an in vitro model of intestinal epithelium. Food Res Int 72:208–214

Sun Z, Cade JR, Fregly MJ et al (1999) Beta-casomorphin induces Fos-like immunoreactivity in discrete brain regions relevant to schizophrenia and autism. Autism 3(1):67–823

Sun Z, Zhang Z, Wang X et al (2003) Relation of β-casomorphin to apnea in sudden infant death syndrome. Peptides 24(6):937–943

Tailford KA, Berry CL, Thomas AC et al (2003) A casein variant in cow's milk is atherogenic. Atherosclerosis 170(1):13–19

Takahashi M, Moriguchi S, Yoshikawa M et al (1994) Isolation and characterization of oryzatensin: a novel bioactive peptide with ileum-contracting and immunomodulating activities derived from rice albumin. Biochem Mol Biol Int 33(6):1151–1158

Takahashi M, Moriguchi S, Suganuma H et al (1997) Identification of casoxin c, an ileum-contracting peptide derived from bovine kappa-casein, as an agonist for c3a receptors. Peptides 18(3):329–336

Tani F, Iio K, Chiba H et al (1990) Isolation and characterization of opioid antagonist peptides derived from human lactoferrin. Agric Biol Chem 54(7):1803–1810

Teschemacher H (2003) Opioid receptor ligands derived from food proteins. Curr Pharm Des 9 (16):1331–1344

Teschemacher H, Koch G, Brantl V (1997) Milk protein-derived opioid receptor ligands. Biopolymers 43(2):99–117

Udenigwe CC, Adebiyi AP, Doyen A et al (2012) Low molecular weight flaxseed protein-derived arginine-containing peptides reduced blood pressure of spontaneously hypertensive rats faster than amino acid form of arginine and native flaxseed protein. Food Chem 132:468–475

Umbach M, Teschemacher H, Praetorius K et al (1985) Demonstration of a beta-casomorphin immunoreactive material in the plasma of newborn calves after milk intake. Regul Pept 12 (3):223–230

Virtanen SM, Laara E, Hypponen E et al (2000) Cow's milk consumption and the childhood diabetes in Finland study group, HLADQB1 genotype, and type 1 diabetes. Diabetes 49:912–917

Waldhoer M, Bartlett SE, Whistler JL (2004) Opioid receptors. Annu Rev Biochem 73(1):953–990

Walther B, Sieber R (2011) Bioactive proteins and peptides in foods. Int J Vitam Nutr Res 81:181–191

Wang J, Hogenkamp DJ, Tran M et al (2006) Reversible lipidization for the oral delivery of leuenkephalin. J Drug Target 14(3):127–136

Wittert G, Hope P, Pyle D (1996) Tissue distribution of opioid receptor gene expression in the rat. Biochem Biophys Res Commun 218:877–881

Yang S, Yunden J, Sonoda S et al (2001) Rubiscolin, a δ selective opioid peptide derived from plant RuBisCo. FEBS Lett 509(2):213–217

Yang S, Kawamura Y, Yoshikawa M (2003) Effect of rubiscolin, a δ opioid peptide derived from RuBisCo, on memory consolidation. Peptides 24(2):325–328

Yoshikawa M (2015) Bioactive peptides derived from natural proteins with respect to diversity of their receptors and physiological effects. Peptides 72:208–225

Yoshikawa M, Tani F, Shiota H et al (1994) Casoxin D, an opioid antagonist ileum-contracting/ vasorelaxing peptide derived from human αs1-casein. In: Brantl V, Teschemacher H (eds) β-Casomorphins and related peptides: recent developments. Wiley, Weinheim, pp 43–48

Yoshikawa M, Takahashi M, Yang S (2003) Delta opioid peptides derived from plant proteins. Curr Pharm Des 9(16):1325–1330

Zioudrou C, Streaty RA, Klee WA (1979) Opioid peptides derived from food proteins. J Biol Chem 254(7):2446–2449

Abstract

Casomorphins are bovine and human milk-derived opioid peptides. These are released from casein milk proteins. Among casomorphins, β-casomorphins (BCMs) are the most abundant group of exorphins released from the β-casein of human and bovine milk. Casomorphins differ from each other on the source of milk (bovine/human) and the number and sequence of amino acids. The commonly studied casomorphins include bovine BCM-4, 5, 6, 7, 8 and neocasomorphin-6 and human BCM-4, 5, 7, 8 and α-casomorphin. Among all casomorphins, BCM-7 is the most significant exorphins studied, because it correlates with the incidence of few human illnesses. BCMs have selectivity for all three types of opioid receptors including μ, δ, and κ. These exorphins (except α-casomorphin) act as agonists for all these opioid receptors. Casomorphins are released from the milk casein during gastrointestinal digestion (in vivo), simulated gastrointestinal digestion (in vitro), or fermentation. Reports have established the production of these peptides from various types of milk, cheese samples, infant formulas, and yogurt. The A1/A2 milk hypothesis establishes that BCM-7 is released from only A1 milk of bovines due to the presence of histidine at position 67 of the β-casein. The presence of proline at this position in A2 milk resists such cleavage and hence this exorphin is not produced from this variant of β-casein.

2.1 Introduction

The human foods like rice, wheat, milk, fish, poultry, and eggs are a rich source of proteins. Milk and eggs provide complete proteins and are very significant from a nutritional perspective. Milk among these dietary elements gains tremendous importance, provides a complete diet for infants and part of the food for adults also. Milk is composed of caseins that are further classified into α, β, and κ and whey proteins that

M. R. Ul Haq, *Opioid Food Peptides*, https://doi.org/10.1007/978-981-15-6102-3_2

are divided into lactalbumin, lactoglobulin, lactoferrin, immunoglobins, etc. During digestion, these proteins yield numerous peptides, which are physiologically active. β-Casein of cow and human milk is a source of β-casomorphins (BCMs). Casomorphins are also released from human milk α-casein and neo-casomorphin-6 is produced from the bovine milk casein. Casoxin A, B, and C are produced from κ-casein of milk on gastrointestinal digestion. The whey proteins α-lactalbumin, β-lactoglobulin, and lactoferrin release α-lactorphins, β-lactorphins, and lactoferroxins, respectively, on gastrointestinal digestion (Teschemacher et al. 1997; Meisel and FitzGerald 2000; Teschemacher and Koch 1991). The human and cow milk-derived peptides demonstrate opioid activity (Zioudrou et al. 1979; Brantl et al. 1979, 1981; Brantl 1984; Lottspeich et al. 1980; Henschen et al. 1979; Brantl and Teschemacher 1979) and morphine-like substance (200–500 ng/l) was isolated from milk (Hazum et al. 1981). The structure and classification of BCMs is the topic of the present chapter. Moreover, the opioid activity and production of BCMs will also be discussed in detail.

2.2   Structure of Casomorphins

BCMs belong to the group of BAPs that demonstrate opioid activity and these are released from the bovine β-casein (De Noni and Cattaneo 2010; Raies et al. 2015). BCM-7 among all these peptides gained more attention because it correlated with human complications. These opioid peptides are composed of 4–11 amino acid residues (Kaminski et al. 2007) and share the first three amino acid residues including tyrosine, proline, and phenylalanine (Muehlenkamp and Warthesen 1996). The most common structural feature of these milk-derived BCMs is the presence of tyrosine amino acid residue at the amino-terminal followed by the occurrence of additional aromatic amino acid residue, phenylalanine or tyrosine, at the 3rd or 4th position that confirms the affinity of this exorphin for the opioid receptors. The demonstration of the opioid activity further depends on the confined negative perspective in the area of the phenolic hydroxyl group of tyrosine, as the elimination of this tyrosine amino acid residue leads to a lack of opioid activity (Kostyra et al. 2004). The structures of BCMs (a-l) are given in Fig. 2.1.

The primary structure of all the bovine BCMs starts from position 60–68 (BCM-8) and is derived from the β-casein of cow milk. BCM-8 is unique, as it possesses histidine or proline at the carboxyl-terminal. BCM-8 shows histidine at the carboxyl end if released from A1 β-casein and proline when produced from A2 β-casein of cow milk. Bovine BCMs (4–7) have the same sequence up to 4th amino acid from the amino-terminal and the other members are added with additional amino acid residues from the carboxyl end. These peptides are released from A1 milk in the gut and have been demonstrated to show opioid activity. Human BCMs (BCM-4, 5, 7, and 8) are also derived from β-casein of human milk. The first three amino acid residues (tyrosine, proline, and phenylalanine) from the amino-terminal are shared between human and bovine BCMs. However, they differ in the sequence of the rest of the amino acids. Human milk-derived BCMs possess a sequence

(a)	Bovine BCM-4 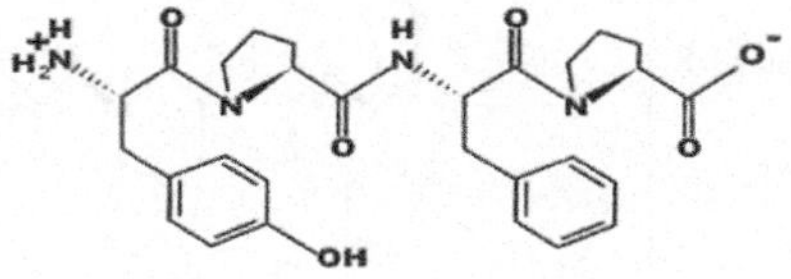
(b)	Bovine BCM-5 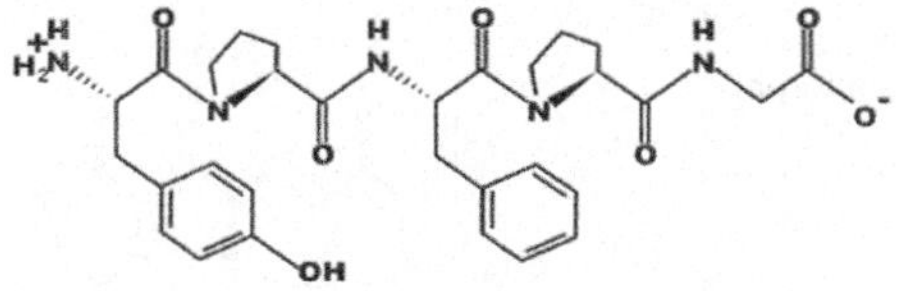
(c)	Bovine BCM-6 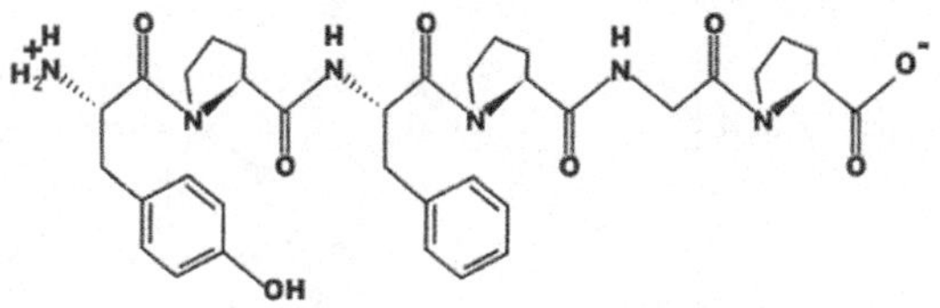
(d)	Bovine BCM-7 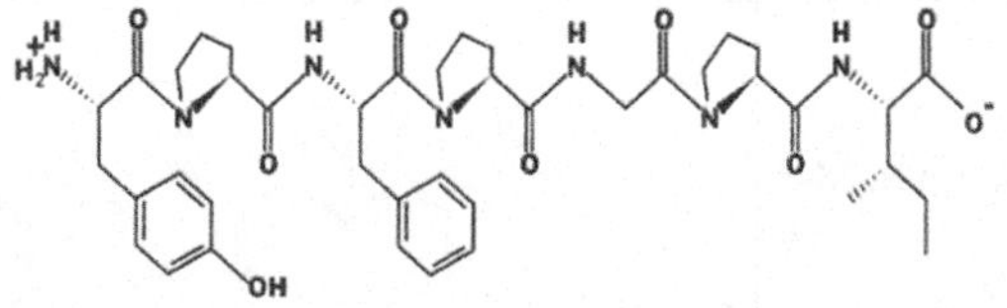
(e)	Bovine BCM-8 (histidine) (A1 beta-casein) 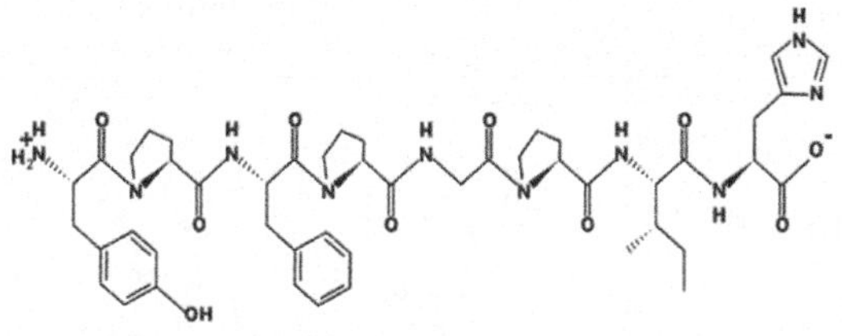

Fig. 2.1 Structures of bovine and human casomorphins

(f)	Bovine BCM-8 (proline) (A2 beta-casein)	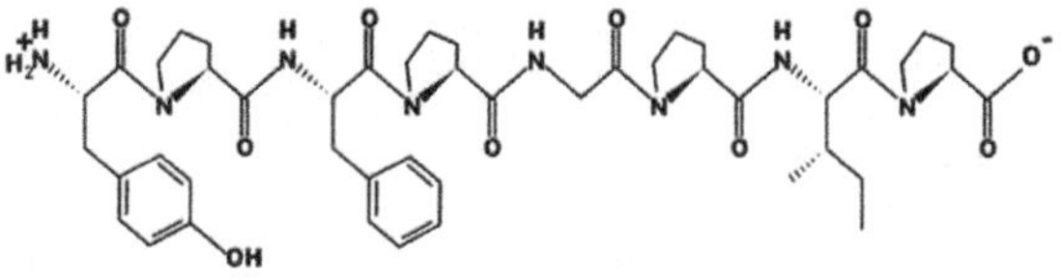
(g)	Bovine neocasomorphin-6	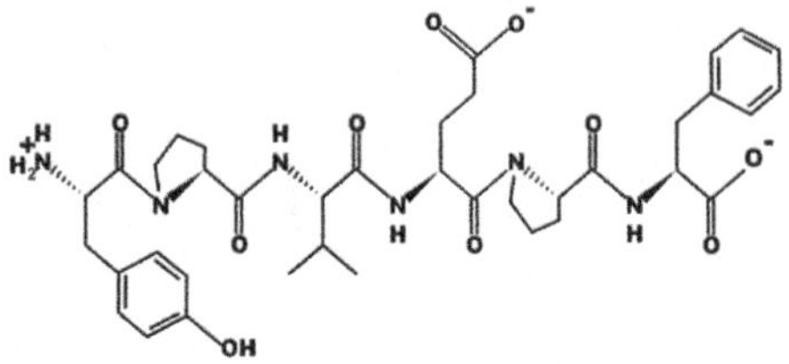
(h)	Human BCM-4	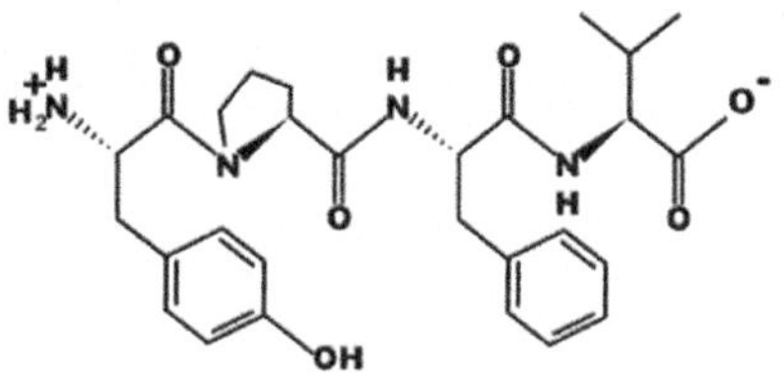
(i)	Human BCM-5	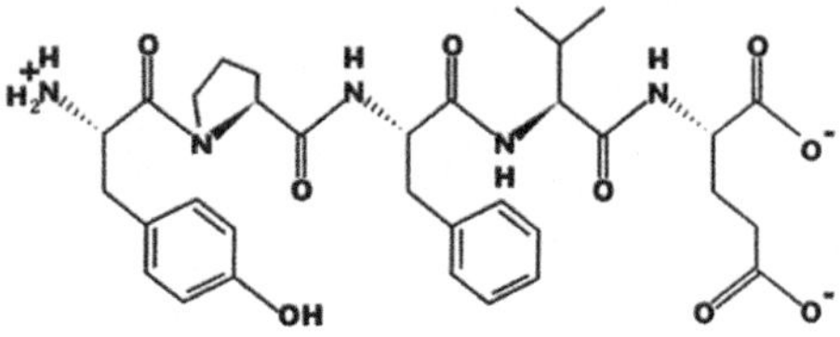
(k)	Human BCM-8	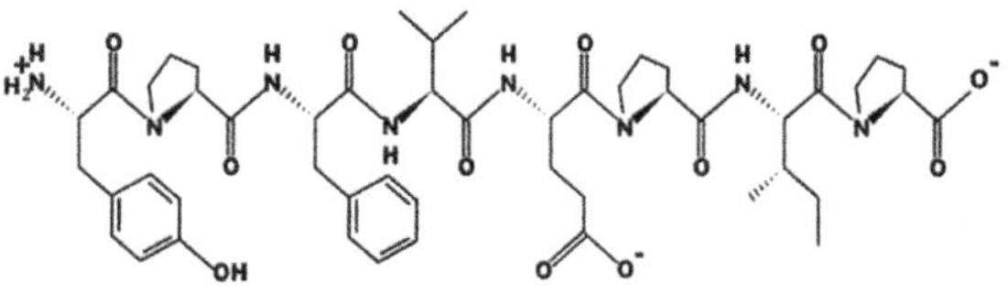
(l)	Human alpha casomorphin	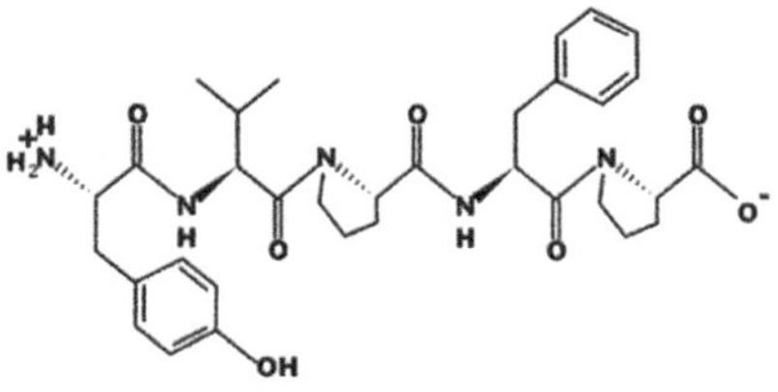

Fig. 2.1 (continued)

(VEPI) that is different from the bovine BCM sequence (PGPI). Human BCMs are found between the positions 51 and 58 of β-casein of milk protein. Moreover, another six amino acid BCM opioid peptide (neocasomorphin-6) is released with a sequence (YPVEPF) from position 114–119 of bovine β-casein. α-Casomorphin is also released from the α-casein of human milk with a sequence YVPFP from positions 158 to 162.

Bovine BCMs (4–7) act as opioid agonists for the μ-opioid receptors. Although bovine BCM-8 also acts as an opioid agonist, it demonstrates this affinity for all three types of opioid receptors (α, κ, and δ). The strength is the highest for the μ-opioid receptor followed by δ- and κ-opioid receptors. Similarly, the bovine neocasomorphin-6 establishes agonistic activity for μ-opioid receptors. Human BCMs (4, 5, 7 and 8) act as agonists for all three types of opioid receptors (α, κ, and δ) and demonstrate the same pattern of affinity for their receptors as denoted by bovine BCM-8. More interestingly, the human casomorphin released from the α-casein acts as agonist or antagonist for all the three types of opioid receptors (α, κ, and δ). However, the strength of the affinity is much stronger for the κ-opioid receptor compared to μ or δ opioid receptors. The primary structure of all α- and β-casomorphins of human and bovine origin is given in Table 2.1.

2.3 Classification of Casomorphins

BCMs are opioid peptides released from human and bovine α- and β-casein of milk. These exorphins are released through gastrointestinal digestion with a combination of various enzymes including pepsin, trypsin, chymotrypsin, elastase, leucine aminopeptidase, elastase, and carboxypeptidase. Although there is no standardized classification of BCMs, these may be classified based on the source of milk and the number of amino acids in these peptides as depicted in Table 2.2.

2.3.1 Based on Chain Length

Based on the number of amino acid residues, BCMs may be BCM-4 to BCM-8, the alphabets refer to the "β-casein" and the numeric part may be the number of amino acids. Bovine β-casein generates BCM-4 to BCM-8, however, in the humans BCM-6 isn't produced. Another BCM, i.e., neocasomorphin-6 composed of 6 amino acids is formed from bovine β-casein, however, its composition is different from the bovine BCM-6. Moreover, human milk gives rise to a unique BCM fragment called α-casomorphin from α-casein that is not formed from the bovine milk. Among bovine BCMs, BCM-5 is the most potent and BCM-7 the least and in between is BCM-4 and BCM-8. However, BCM-7 is the most widely studied bovine BCM because it correlates with various human illnesses.

Table 2.1 Primary structure or sequences of milk-derived casomorphins (modified from Garg et al. 2016)

Casomorphins	Primary structure
Bovine BCM-4	Nt. . . .YPFP. . . . . .Ct
Bovine BCM-5	Nt. . . .YPFPG. . . . . .Ct
Bovine BCM-6	Nt. . . .YPFPGP. . . . . . . .Ct
Bovine BCM-7	Nt. . . .YPFPGPI. . . . . . . .Ct
Bovine BCM-8	Nt. . . .YPFPGPI.H/P. . . . ..Ct
Bovine neocasomorphin-6	Nt. . . .YPVEPF. . . . . . . .Ct
Human BCM-4	Nt. . . .YPFV. . . . . .Ct
Human BCM-5	Nt. . . .YPFVE. . . . . .Ct
Human BCM-7	Nt. . . .YPFVEPI. . . . . .Ct
Human BCM-8	Nt. . . .YPFVEPIP. . . . . . . .Ct
Human α-casomorphin	Nt. . . .YVPFP. . . . . .Ct

Nt amino (N) terminal, *Ct* carboxyl (C) terminal

2.3.2 Based on Milk Source

Depending upon the source of milk, BCMs may be categorized into bovine and human BCMs. Bovine milk on proteolysis yields BCM-4 to BCM-8. Similarly, in humans the same peptides are formed, nevertheless, BCM-6 is absent. The first three amino acids in BCMs (tyrosine, proline, and phenylalanine) are shared in humans and bovines, the rest differ in both types of milk. Additionally, neocasomorphin-6 is formed from bovine β-casein and α-casomorphin from human milk. Much of the importance has been paid towards the bovine BCMs because they correlated with human illnesses. However, both the types of BCMs have their agonistic or antagonistic affinities with μ, δ, and κ-opioid receptors. These peptides are responsible for the manifestation of various physiological actions that is the subject of Chap. 3.

2.4   Production of Casomorphins

Opioid peptides are released from milk during gastrointestinal digestion, fermentation, and under simulated gastrointestinal digestion (SGID) in vitro using various combinations of enzymes. Recently, in 2015 we studied in our laboratory, the release of BCMs from the milk of crossbred cattle Karan Fries. We first screened the crossbred cattle for A1 and A2 alleles of β-casein by PCR-RFLP. After genotyping, milk was collected from these cattle which were regarded as A1A1 homozygous β-casein, A1A2 heterozygous β-casein, and A2A2 homozygous β-casein. SGID was first carried out using pepsin, trypsin, and chymotrypsin and the analysis was carried out at analytical RP-HPLC. The commercially synthesized BCM-5 and BCM-7 served as standards for the analysis of analytical RP-HPLC. The two chemically synthesized standard peptides BCM-5 and BCM-7 demonstrated retention times at 24 and 28 min, respectively. The hydrolysates were run on analytical RP-HPLC for the demonstration of peaks at these retention times. The peptide fractions were collected corresponding to these retention times on preparative HPLC and then

Table. 2.2 Classification of casomorphins

Classification on chain length		Classification on milk source	
		Bovine	Human
BCM-4	Sequence	YPFP	YPFV
	Receptor selectivity	μ	$\mu > \delta > \kappa$
	Agonist/ antagonist	A	A
	References	Brantl et al. (1981)	Chiba et al. (1989)
BCM-5	Sequence	YPFPG	YPFVE
	Receptor selectivity	μ	$\mu > \delta > \kappa$
	Agonist/ antagonist	A	A
	References	Brantl et al. (1981)	Teschemacher et al. (1997)
BCM-6	Sequence	YPFPGP	Not released
	Receptor selectivity	μ	–
	Agonist/ antagonist	A	–
	References	Brantl et al. (1981)	–
BCM-7	Sequence	YPFPGPI	YPFVEPI
	Receptor selectivity	μ	$\mu > \delta > \kappa$
	Agonist/ antagonist	A	A
	References	Brantl et al. (1981)	Teschemacher et al. (1997)
BCM-8	Sequence	YPFPGPI.H/P	YPFVEPIP
	Receptor selectivity	$\mu > \delta >> \kappa$	$\mu > \delta > \kappa$
	Agonist/ antagonist	A	A
	References	Teschemacher et al. (1997)	Teschemacher et al. (1997)
Neocasomorphin-6	Sequence	YPVEPF	Not released
	Receptor selectivity	μ	–
	Agonist/ antagonist	A	–
	References	Jinsmaa and Yoshikawa (1999)	–
α-Casomorphin	Sequence	Not released	YVPFP
	Receptor selectivity	–	$\mu/\delta <<< \kappa$
	Agonist/ antagonist	–	A/AN
	References	–	Antila et al. (1991)

freeze lyophilized. After MS-MS analysis, none of the peaks showed the presence of either BCM-5 or BCM-7. However, a precursor peptide sequence of 14 amino acid residues was obtained. When this fraction with precursor sequence was further digested with the elastase and leucine aminopeptidase, a seven amino acid peptide from A1A1 and A1A2 milk was assessed through competitive ELISA and isolated organ bath. BCM-5 was not recovered from any of the milk variants. The opioid activity of the generated peptides was checked on rat ileum from homozygous A1 variant was 0.534–0.595 μM and that of the heterozygous variant was 0.410–0.420 μM that, in turn, validated the presence of BCM-7 (Raies et al. 2015). Earlier the researchers tried to evaluate the presence of BCM-7 and related peptides in milk, cheese, yogurt, and other dairy products while exploiting various bioanalytical techniques. The reverse phase-HPLC was the main technique employed for this purpose. This technique, however, suffers the problem that different peptides with the same hydrophobicity index elute at the same retention time. The enzyme-linked immunosorbent assays (ELISA) could detect these peptides quantitatively; however, these may show cross-reactivity and may not be able to detect exact peptide sequences. The opioid activity determining assays could assess the opioid activity on animal tissues like GPI but are insufficient in determining the sequences. More recently many researchers have tried the RP-HPLC coupled with MS-MS for quantitative determination and sequence elucidation. Therefore these combined analytical methods are the best approaches for peptide identification and quantification. The investigators in 2007 detected the presence of BCM-7 in raw milk samples, however, the accurate regards of the presence of somatic cells in the milk was taken into consideration (Cieslinska et al. 2007). Therefore, EFSA in 2009 commented against such analysis and argued that the production of BCM-7 in such raw milk may be due to the presence of proteolytic enzymes released from somatic or microbial cells (European Food Safety Authority (EFSA) 2009). Contrariwise to the EFSA's arguments, Napoli and coworkers had earlier demonstrated that this peptide is not generated in cow milk that had contracted mastitis and these cells (somatic) were greater than 500,000/mL (Napoli et al. 2007). Moreover, the same researcher again in 2012 detected the presence of BCM-7 in raw milk of healthy cows and the highest yield was from A1A1, followed by heterozygous A1A2 and A2A2 homozygous A2A2 cows (Cieslinska et al. 2007). Therefore, these contradictions regarding the release of BCM-7 in raw milk deserve further investigation. Accurate regards should be taken for the detection of these peptides from raw milk of healthy and unhealthy animals in addition to the presence of somatic and bacterial cells. Moreover, the release of these peptides in human milk may help in understanding the production of the same types of peptides in earlier studies performed in bovine milk. It was Jarmolowska and coworkers who assessed the higher yield of BCM in colostrum compared to the mature milk. These investigators collected the milk samples from healthy women and immediately stored them at −70 °C, therefore, strategy excluded all the probabilities of the generation of these opioid peptides with somatic and bacterial-derived proteases (Jarmolowska et al. 2007). This perspective may lead to the conclusion of the production of BCMs from the milk by endogenous proteases of mammary glands (Koch et al. 1985). The changes that occur in proteins of the infant formulas during sterilization may play a role in the hydrolysis and turn the release of BAPs. Researchers while studying SGID have established that glycation intensely hindered proteolysis, nevertheless, the cross-linking that occurs through β-elimination reactions has a non-significant effect. Among many BCMs analyzed, only BCM-7 was assessed at a concentration of 0.187–0.858 mg/l. The concentration of BCM-7 recovered from sterilized model systems that were manufactured without the use of any sugars varied from 0.256 to 0.655 mg/l. The production of BCM-7 during gastrointestinal was delayed by glycation (Cattaneo et al. 2017). Similarly in a clinical trial performed in healthy humans, the researchers found that large and medium-sized peptides were generated after 3 and 6 h, respectively, from casein and whey proteins. They further found that the β-casein was the predominant protein precursor for the generation of various physiologically active peptides. The quantity of BCMs (57-, 58-, 59-, 60–66) and β-casein fragments (108–113) produced after food consumption was enough to stimulate the physiological actions including antihypertensive and opioid activity (Boutrou et al. 2013). The stability of BCM-7 in the intestine during its formation was assessed in vitro while exploiting the activities of brush border proteases (BBM, porcine) and gastrointestinal juices (GI, humans). The chemically synthesized BCM-7 was allowed to incubate with these two sources of enzymes. The investigators found that this peptide is partially hydrolyzed with GI enzymes. Regarding the use of BBM, 42% and 79% of this peptide were degraded after 2 h and 4 h digestion, respectively. Therefore, the authors concluded that this peptide was hydrolyzed with GI and BBM digestion, nevertheless, a small fraction (5%) was still identified after 1-day incubation. However, it was unclear whether the intact amount of peptide was transported through active transporters in the gut epithelial cells and reached the blood circulation (Asledottir et al. 2019).

There are some reports available that show the release of BCMs in various cheese varieties (Sienkiewicz-Szłapkaa et al. 2009; De Noni and Cattaneo 2010). The amount of BCM-7 assessed in Brie cheese was 5–15 mg/g (Jarmolowska et al. 1999), 0.15 mg/g (De Noni and Cattaneo 2010), and 6.48 mg/g (Sienkiewicz-Szłapkaa et al. 2009). However, De Noni and Cattaneo detected 0.01–0.11 mg/g in different cheese samples (Gouda, Gorgonzola, Cheddar, and Fontina) (De Noni and Cattaneo 2010). Similarly, the levels of BCM-7 released from semi-hard cheeses were lower compared to Polish mold cheeses. Moreover, BCM-5 was also found in some other cheese samples (Sienkiewicz-Szłapkaa et al. 2009). Moreover, other researchers detected the release of BCM-9 and BCM-10 in Gouda (Toelstede and Hofmann 2008) and BCM-11 in Italian goat cheese (Rizzello et al. 2005). Earlier Muehlenkamp and Warthesen could not detect BCM-7 in Cheddar and Brie cheese samples. The investigators employed RP-HPLC for analysis. The researchers speculated that either the formation of BCM-7 from these cheese samples was unlikely or the formation was followed by degradation during ripening. Moreover, the released peptides may be produced in the amounts below the sensitivity levels of the RP-HPLC. Furthermore, the production of BCMs from cheeses may be due to the secretion of enzymes by starter culture (Muehlenkamp and Warthesen 1996). In 2010, De Noni and Cattaneo recovered BCM-7 (0.11 mg/g) from cheddar cheese.

Almost 10 years before the former investigation, other researchers argued that this opioid peptide is not the original milk-derived peptide, because the elimination of these protein fragments occurs through whey while curd formation (Jarmolowska et al. 1999). The aforementioned type of cheese is manufactured through the employment of starter cultures (*Lactococcus. lactis* ssp. *Cremoris* and *L. lactis ssp. Lactis*) (Robinson 1995). In this perspective, some of the researchers assume that proteases secreted from above-mentioned bacteria have potential to reduce the quantity of this opioid peptide by half at certain conditions (1.5% NaCl and pH 5.0) after 6–15 weeks; these criteria exactly match in the preparation of Cheddar cheese (Muehlenkamp and Warthesen 1996). Other investigators assumed that the levels of this opioid peptide may be high in the beginning compared to the ripened stage (De Noni and Cattaneo 2010). This may be because initially this peptide is formed on the addition of starter culture from β-casein but later degraded during ripening. Besides starter cultures, there is a possibility of production of this peptide in cheese due to the presence of endogenous proteolytic enzymes and nonstarter culture peptidases obtained from contaminated equipment, udder, and pipes (European Food Safety Authority (EFSA) 2009). On the contrary, Gaucher and coworkers could not recover BCMs while utilizing plasmin (endogenous protease) and rennet enzymes employed for milk coagulation (Gaucher et al. 2008; McSweeney 2004). It was Hayaloglu and coworkers in 2010 who worked on peptide profiling in Malatya cheese that was manufactured from raw milk where no use of starter cultures was performed. These researchers although established the presence of various peptides in the cheese but did not investigate BCMs. When the authors compared peptide profiling between raw cheese and pasteurized cheese, they argued that resultant peptides may have formed from the proteases of nonstarter culture (Hayaloglu et al. 2010). It is also possible that the generation of BCM-7 in the cheese is due to the proteolytic activity of molds used in cheese ripening. The amount of BCM-7 in Brie cheese (De Noni and Cattaneo 2010) and Rokpol cheese (used fungi and LAB) (Sienkiewicz-Szłapkaa et al. 2009) was significantly higher compared to the cheese employed LAB only. When the analysis of the BCM-7 was performed in cheese samples through purified enzymes like Neutrase isolated from *B. subtilis*. The presence of BCM-7 was detected but not quantified (Haileselassie et al. 1999). This opioid peptide was not detected in samples that were supplemented with Neutrase and crude extract of *L. casei* and samples added with Neutrase and a peptidase (from *Lactococcus lactic* and *Aspergillus oryzae*). The authors concluded that the existence of the BCM-7 in cheddar cheese may be due to the proteolysis of β-casein by Neutrase. The subsequent absence of this peptide may be due to hydrolysis by proteolytic enzymes up to the levels not detected by bioanalytical techniques. This, therefore, implies the production and degradation of BCM-7 in the samples by proteolytic enzymes of starter cultures. Overall it leads to the conclusion that the presence of large BCM precursors helps us comprehend the processing utilized in BCM formation. More generally, the proteolytic enzymes hydrolyze milk β-casein to large fragments including BCM precursors, and consequently few precursors are changed into BCMs by specific proteolytic enzymes (Haileselassie et al. (1999). The BCM precursor sequences (f58-72) from milk β-casein were identified in Crescenza, Cheddar, and Jarlsberg cheese when allowed to ripen for 8–10 days (Smacchi and Gobbetti 1998; Stepaniak et al. 1995). This precursor sequence has a probability to degrade into BCM-13 as reported by Jinsmaa and Yoshikawa (1999) when stored for a longer time. The aforementioned studies evaluated the presence of BCM sequences from bovine milk irrespective of genetic variants of β-casein. More recently, the debate is on A1 and A2 "like" milk. The A1 milk that has histidine at position 67 allows the proteolytic cleavage and is, therefore, a source of BCM-7; A2 milk has proline at this position and resists such cleavage. This perspective has been ignored by all the investigators except De Noni and Cattaneo (2010) who kept accurate regard for genetic variants of β-casein. However, the authors were uncertain which of the genetic variants of β-casein lead to the generation of BCM-7 after fermentation (De Noni and Cattaneo 2010). In conclusion, to elucidate the production of BCM-7 from the cheese, it will be prerequisite that milk should be screened for A1 and A2 variants of β-casein.

In 2007, the investigators purified peptides of β-casein origin from yogurt preincubated with probiotic bacteria (*L. acidophilus* L10, *L. casei* L26 and *B. lactis* B94) and a yogurt starter culture. These researchers could neither isolate intact BCMs nor their precursors (Donkor et al. 2007). Similarly, other researchers isolated at least thirty peptides from yogurt that was fermented with *S. salivarius* ssp. *Thermophilus* and *L. delbrueckii* ssp. *Bulgaricus*. These peptides were of A1 β-casein origin and precursors of BCM including f57–72 and f57–68 containing histidine at 67th position of the parent β-casein. These observations, therefore, point out the explanation that BCMs are more likely to be generated from A1 "like" variants of β-casein (Schieber and Brückner 2000). Moreover, numerous peptide fragments were isolated from the alternate variants of β-casein (A1 and A2) during the preparation of yogurt. Among these fragments, a precursor of BCM-7 with f58–72 was identified from β-casein. Additionally, other fragments like f59–70 and f59–68 were generated from A2 genetic variant of β-casein and there is a great possibility of these precursors to get broken down to BCM-9, BCM-10, and BCM-13 (Kunda et al. 2012). The HPLC coupled with high-resolution mass spectrometry was employed for the study of the stability of BCM-5 and BCM-7 by *Lactobacillus delbrueckii* ssp. *Bulgaricus* and *Streptococcus thermophilus* and to analyze the peptide degradation products generated during yogurt processing. The ultra-heat treated bovine milk was fermented with a mixture of aforesaid strains and with each strain also. The pH was adjusted to 4.5 and storage was performed for 1–7 days at 4 °C. The investigators found that both bacterial strains degraded the BCMs (5 and 7) on fermentation and the degree of proteolysis depended both on the strain used and storage time. However, it could not be assessed whether degradation products resulted from BCMs or intact proteins. Nevertheless, the production of a dipeptide (YP) and tripeptide (GPI), the parts of sequences of BCMs was assessed through involvement of *L. delbrueckii* ssp. *bulgaricus*. Therefore it implies that there is a probability for the degradation of BCMs through this strain (Nguyen et al. 2018). In the future, it seems that these different strains of lactobacilli may have enough potential for further use in food safety and management. Since the source of degraded di and tripeptide was not established, it warrants more research in this research area. Moreover, studies have established that A1 genetic variant of β-casein of yogurt yielded f60–72 that could be a source of BCMs. Many dipeptides have been purified in yogurt (f60–61 and f62–63) and it was assumed that these are produced from both the genetic variants of β-casein. It was therefore concluded that the proteases released from bacteria in yogurt cleave peptide bonds between proline and another amino acid that may cause the breakdown of BCMs. The cleavage of BCMs in the processing of yogurt utilizes dipeptidases which are in turn secreted from LAB and cleave linkages between proline and other amino acids (Gobbetti et al. 2002). Similarly, Farvin and coworkers purified peptides from yogurt (commercial) and most of the peptides were generated from β-casein, however, few peptides were released from α- and κ-casein also. BCM-9 but not BCM-7 was recovered from the A2 genetic variant of β-casein (Farvin et al. 2010). However, both the research groups (Kunda, Farvin) did not characterize the starter culture employed in yogurt that may be one of the main elements for checking the formation and degradation of BCMs. Although De Noni and Cattaneo kept accurate regard of the genetic variants of β-casein and identified alternate variants of β-casein (A1/A2/ B) in their study, they still could not identify BCM-7 opioid peptides (De Noni and Cattaneo 2010). Similar observations were reported by other investigators in the yogurt samples (21 days at 4 °C). Therefore these studies lead to the conclusion that the starter culture either alone or a mixture is unable to release BCMs from the milk proteins. Already, it was demonstrated that BCM-7 is not recovered from Taleggio cheese through the involvement of *L. delbrueckii* ssp. *bulgaricus* and *S. thermophiles* as a starter culture. The Taleggio and yogurt used in the studies were stored for almost 1 month (De Noni and Cattaneo 2010).

It is also possible that the cytoplasmic proteases like PepX take more time for secretion and hydrolyze proteins to yield BCMs (Atlan et al. 1990). Additionally, other investigators believe that the proteolytic enzymes linked to the starter culture or other contaminating microorganisms may find it easy to cleave caseins into large protein a fragment including BCM-7, the latter is further hydrolyzed by DPPIV to di and tri-peptides (Nielsen et al. 2009). Moreover, the bacteria (*S. thermophilus* and *L. delbrueckii ssp. Bulgaricus*) used for the preparation of yogurt may cleave proteins into BCM-7 and other peptides which may further be degraded into smaller peptides. The formation and degradation of BCM and related peptides are strongly under the control of the symbiotic association of the above-mentioned bacteria. It is well established that the growth of either is dependent on each other, the products of one bacterium stimulates other and vice versa released by each organism stimulate the growth of the other (Tamime and Robinson 1999). As an instance, the protein fragments and individual amino acids generated through *L. delbrueckii ssp. Bulgaricus* encourage the growth of *S. thermophilus* (Courtin et al. 2002). From this perspective, it is suggested that BCMs may be produced from the former and degraded by the latter. However, other workers have established that a single strain of the bacterium (*L. delbrueckii ssp. bulgaricus SS1*) used for yogurt fermentation generated many peptide fractions from β-casein; however, BCMs were not recovered (Gobbetti et al. 2000). Similarly, other investigators purified peptide fragments (f176–180 and f1–8) from the β-casein protein of milk in yogurt using *L. delbrueckii ssp. Bulgaricus.* To sum up, it is concluded that it is not clear when and how these BCMs are generated and hydrolyzed in yogurt while using starter cultures like *S. thermophilus* and *L. delbrueckii ssp. Bulgaricus* (Papadimitriou et al. 2007). Therefore well-designed experiments are needed to explore in this research area and demonstrate the stability of the BCMs. Moreover, genetic variants of β-casein should be given special consideration.

2.5 Effect of Milk Processing on Casomorphin Production

The heat treatment plays a significant role in the dairy sector from food safety and quality perspective. It destroys in milk the disease and spoilage-causing bacteria in addition to the benefial effect for the introduction of viscosity and texture to the fermented milk products (Krasaekoopt 2003; Hattem et al. 2011). When yogurt is manufactured in the dairy industry, pasteurization is often followed by the secondary heating at 90–95 °C for 5 min (European Food Safety Authority (EFSA) 2009). This treatment leads to the denaturation of proteins (whey) and increases interaction between β-lactoglobulin and κ-casein. The establishment of this aggregate affects the protein hydrolysis during fermentation (El-Zahar et al. 2003). Moreover, the increased temperature increases the protein degradation of intact protein to peptides (Gaucheron et al. 1999). The peptides are not only generated through enzymatic but also through increased temperatures (Lotfi 2004; Meltretter et al. 2008). The researchers identified five novel protein fragments when milk was heated to 120 °C for 30 min, these peptides were not initially present in raw milk. Using a more sophistical analytical approach (MALDI-TOF-MS), these peptides were of the casein origin; however, they did not match with BCM sequences. Therefore, it was concluded that simple heat treatment could not release BCM sequences from milk. At the same time when the temperature of the milk is increased, it leads to the inactivation of enzymes thereby the chances of the change of pH decrease and in turn, the peptide formation by acid and proteases may not be taken into consideration. Moreover at high temperatures the rate of Maillard's reaction enhances resulting in the production of free radicals, the latter attacks the peptide bonds to form peptides (Meltretter et al. 2008). The generation of new protein fragments in milk after heat treatment depends on the thermal stability of peptide bonds of β-casein of milk protein. Researchers have reported the formation of ten peptides in raw milk when heated to 120 °C for half an hour. Among these ten peptides, two peptides were produced from β-casein and none of the sequences resembled the BCMs or their precursors (Gaucheron et al. 1999). When milk was given high heat treatments and stored for 6 months at 20 °C, pro-BCMs were observed through cathepsins and elastase (Gaucher et al. 2008). A protein fragment (f57–68) corresponding to β-casein was detected in yogurt that is assumed to have generated due to heat treatment, the latter is an essential process during yogurt preparation (Schieber and Brückner 2000). While other researchers kept accurate regard of A1/A2 genetic variants of milk β-casein and tried to assess BCM-7 in these types of milk. Commercially pasteurized, ultra-heat treated (UHT) and in-bottle heat-treated milk was taken for consideration. The investigators, however, could not quantify BCM-7 in these samples (De Noni and Cattaneo 2010). Cieslinska and coworkers tried to assess BCM-7 in different types of milk containing homozygous A1, homozygous A2, and a mixture of A1 and A2 by employing ELISA. The varieties of milk samples were pasteurized at 95 °C for 20 min and sterilized at 117 °C for 5 min. The researchers found that the amount of this peptide decreased when milk was heated to high temperatures. The authors argue that the peptide degradation occurred not due to direct heat on peptide bonds but because of the interaction between amino acid residues of the BCM-7 and lactose that decreased the affinity of the altered BCM-7 to the antibody employed in the ELISA that further led to underestimated BCM-7 levels (Cieślińska et al. 2012). The cow milk protein hydrolysates were employed for the fortification of pasteurized donor human milk (PDHM) for the application of preterm infant nutrition. The same study was performed to analyze the effect of heat and production of BCMs. Three commercial fortifiers were chosen for analysis of thermal protein modifications through heat load indexes. Before digestion, BCMs (3–7) were present in the fortifiers. After SGID corresponding to that of the premature infant, bovine BCMs were still detected in fortified PDHM and, however, to a surprise human BCMs (3–9) were formed. These results may lead to the conclusion that better demonstrate the nutritional characteristics of protein fortifiers in addition to fortified PDHM anticipated for the dietary purpose of preterm infants (Cattaneo et al. 2020).

As aforesaid many epidemiological followed by few animal and human trials have correlated consumption of opioid peptides like BCMs with various physiological disorders. It is assumed that food proteins on digestion in the gastrointestinal tract release some opioid peptides that have a role in the incidence of these human illnesses. On the other hand, studies have established that Bifidobacterium a class of probiotic bacteria is a major gut microbiota in humans especially infants. In this perspective, the investigators found that these bacteria express dipeptidyl peptidase IV (oligo prolyl peptidase) that has specificity for the proline residues, a major amino acid residue in opioid peptides. These enzymes, therefore, have specificity in cleaving in these opioid peptides to dipeptides and hence relieving their adverse effects. The authors concluded that Bifidobacterium species have a potential for the hydrolysis of food-opioid peptides. These have the potential to be used as probiotics for the elimination of food-derived complications and may contribute to consumer health (Sakurai et al. 2018).

2.6   Opioid Activity of Casomorphins

From the last four decades, the researchers have established that the peptide fragments are released from the intact proteins on digestion in vivo or SGID (in vitro). Many of these peptides have the potential to bind with the opioid receptors like MOP, KOP, and DOP and demonstrate opioid activity. These receptors are located in the central nervous system, peripheral nervous system, gastrointestinal tract, and some immune cells. The opioid activity of milk-derived opioid peptides dates back to 1979 when the opioid activity of BCMs was demonstrated (Brantl et al. 1979, 1981; Zioudrou et al. 1979; Koch et al. 1985). This opioid activity was assessed with the help of standard pharmacological techniques like guinea pig ileum, mouse vas deferens (MVD), longitudinal muscle myenteric plexus preparation (GPI), and ligand-binding assays. The results indicated that both human and bovine BCMs significantly bind with MOP. Moreover, some BCMs like BCM-4 and BCM-5 have an affinity towards DOP (Koch et al. 1985). When comparative studies were performed for the human and bovine BCMs, it was revealed that in GPI assay that human BCMs (Tyr-Pro-Phe-Val-Glu-Pro-Ile) observed 3–30 times less potency compared to bovine BCMs (Tyr-Pro-Phe-Pro-Gly-Pro-Ile) and further 300–600 times less potency compared to normorphine (Koch et al. 1985). The order of potencies correlated well with both the human and bovine BCMs (BCM5 > BCM4 > BCM8 > BCM7). Moreover, it was also found that bovine BCMs have an affinity for opioid receptors found in rat brain when observed through IC_{50} values, the potency of bovine BCMs was almost 300 times less than morphine. BCM-5 was found to be the most active and the BCM-7 the least one (Brantl et al. 1981). We have also checked the opioid activity of BCM-7 in homozygous (A1A1) and heterozygous (A1A2) β-casein hydrolysates. These hydrolysates were prepared with pepsin, trypsin, and chymotrypsin. These two types of hydrolysates were detected for the presence of 14 amino acid residue precursor of BCM by MS-MS. The hydrolysate with precursor was further digested with elastase and leucine aminopeptidase and confirmed for the presence of BCM-7 with competitive ELISA. The confirmed hydrolysates for BCM-7 were checked for the assessment of opioid activity on rat ileum. The observations of our study indicated IC_{50} of 0.534 µM and 0.595 µM for homozygous A1 β-casein hydrolysates with PeTEL and PeTCEL combinations of enzymes, respectively. Moreover, the IC_{50} of heterozygous β-casein hydrolysates was 0.410 µM and 0.420 µM with PeTEL and PeTCEL protease combinations, respectively (Raies et al. 2015).

References

Antila P, Paakkari I, Järvinen A et al (1991) Opioid peptides derived from in-vitro proteolysis of bovine whey proteins. Int Dairy J 1(4):215–229

Asledottir T, Picariello G, Mamone G (2019) Degradation of β-casomorphin-7 through in vitro gastrointestinal and jejunal brush border membrane digestion. J Dairy Sci 102(10):8622–8629

Atlan D, Laloi P, Portalier R (1990) X-prolyl-dipeptidyl aminopeptidase of lactobacillus delbrueckii subsp. bulgaricus: characterization of the enzyme and isolation of deficient mutants. Appl Environ Microbiol 56(7):2174–2179

Boutrou R, Gaudichon C, Dupont D (2013) Sequential release of milk protein-derived bioactive peptides in the jejunum in healthy humans. Am J Clin Nutr 97(6):1314–1323

Brantl V (1984) Novel opioid peptides derived from human beta-casein: human beta-casomorphins. Eur J Pharmacol 106(1):213–214

Brantl V, Teschemacher H (1979) A material with opioid activity in bovine milk and milk products. Naunyn Schmiedebergs Arch Pharmacol 306(3):301–304

Brantl V, Teschemacher H, Henschen A et al (1979) Novel opioid peptides derived from casein (beta-casomorphins). I. Isolation from bovine casein peptone. Hoppe Seylers Z Physiol Chem 360(9):1211–1216

Brantl V, Teschemacher H, Bläsig J et al (1981) Opioid activities of beta-casomorphins. Life Sci 28 (17):1903–1909

Cattaneo S, Stuknytė M, Masotti F et al (2017) Protein breakdown and release of β-casomorphins during in vitro gastro-intestinal digestion of sterilized model systems of liquid infant formula. Food Chem 217:476–482

Cattaneo S, Pica V, Stuknytė M et al (2020) Effect of protein fortification on heat damage and occurrence of β-casomorphins in undigested donor human milk intended for nutrition of preterm infants. Food Chem 314:126176. https://doi.org/10.1016/j.foodchem.2020.126176

Chiba H, Tani F, Yoshikawa M (1989) Opioid antagonist peptides derived from beta-casein. J Dairy Res 56(3):363–366

Cieslinska A, Kaminski S, Kostyra E et al (2007) β-Casomorphin-7 in raw and hydrolyzed milk derived from cows of alternative β-casein genotypes. Milchwissenschaft 62:125–127

Cieślińska A, Kostyra E, Kostyra H et al (2012) Milk from cows of different β-casein genotypes as a source of β-casomorphin-7. Int J Food Sci Nutr 63(4):426–430

Courtin P, Monnet V, Rul F (2002) Cell-wall proteinases PrtS and PrtB have a different role in Streptococcus thermophilus/Lactobacillus bulgaricus mixed cultures in milk. Microbiology 148 (11):3413–3421

De Noni I, Cattaneo S (2010) Occurrence of b-casomorphins 5 and 7 in commercial dairy products and in their digests following in vitro simulated gastro-intestinal digestion. Food Chem 119:560–566

Donkor ON, Henriksson A, Singh TK et al (2007) ACE inhibitory activity of probiotic yoghurt. Int Dairy J 17(11):1321–1331

El-Zahar K, Chobert JM, Sitohy M et al (2003) Proteolytic degradation of ewe milk proteins during fermentation of yoghurts and storage. Food/Nahrung 47(3):199–206

European Food Safety Authority (EFSA) (2009) Review of the potential health impact of β-casomorphins and related peptides. Sci Rep 231:1–107

Farvin KS, Baron CP, Nielsen NS et al (2010) Antioxidant activity of yoghurt peptides: part 2– characterisation of peptide fractions. Food Chem 123(4):1090–1097

Garg S, Nurgali K, Mishra V (2016) Food proteins as source of opioid peptides-A review. Curr Med Chem 23(9):893–910

Gaucher I, Mollé D, Gagnaire V et al (2008) Effects of storage temperature on physico-chemical characteristics of semi-skimmed UHT milk. Food Hydrocoll 22(1):130–143

Gaucheron F, Molle D, Briard V et al (1999) Identification of low molar mass peptides released during sterilization of milk. Int Dairy J 9(8):515–521

Gobbetti M, Ferranti P, Smacchi E et al (2000) Production of angiotensin-I-converting-enzyme-inhibitory peptides in fermented milks started by *lactobacillus delbrueckii subsp. bulgaricus* SS1 and *Lactococcus lactis subsp. cremoris* FT4. Appl Environ Microbiol 66(9):3898–3904

Gobbetti M, Stepaniak L, De Angelis M et al (2002) Latent bioactive peptides in milk proteins: proteolytic activation and significance in dairy processing. Crit Rev Food Sci Nutr 42 (3):223–239

Haileselassie SS, Lee BH, Gibbs BF (1999) Purification and identification of potentially bioactive peptides from enzyme-modified cheese. J Dairy Sci 82(8):1612–1617

Hattem HE, Manal AN, Hanaa SS et al (2011) A study on the effect of thermal treatment on composition and some properties of camel milk. J Brew Distill 2(4):51–55

Hayaloglu A, Deegan K, McSweeney P (2010) Effect of milk pasteurization and curd scalding temperature on proteolysis in Malatya, a Halloumi type cheese. Dairy Sci Technol 90(1):99–109

Hazum E, Sabatka JJ, Chang KJ et al (1981) Morphine in cow and human milk: could dietary morphine constitute a ligand for specific morphine (mu) receptors? Science 213 (4511):1010–1012

Henschen A, Lottspeich F, Brantl V (1979) Novel opioid peptides derived from casein (beta-casomorphins). II. Structure of active components from bovine casein peptone. Hoppe Seylers Z Physiol Chem 360(9):1217–1224

Jarmolowska B, Kostyra E, Krawczuk S et al (1999) β-Casomorphin-7 isolated from brie cheese. J Sci Food Agric 79:1788–1792

Jarmolowska B, Szlapka-Sienkiewicz E, Kostyra E et al (2007) Opioid activity of humana formula for newborns. J Sci Food Agric 87(12):2247–2250

Jinsmaa Y, Yoshikawa M (1999) Enzymatic release of neocasomorphin and β-casomorphin from bovine β-casein. Peptides 20(8):957–962

Kaminski S, Cieoelinska A, Kostyra E (2007) Polymorphism of bovine beta-casein and its potential effect on human health. J Appl Genet 48(3):189–198

Koch G, Wiedemann K, Teschemacher H (1985) Opioid activities of human β-casomorphins. Naunyn Schmiedeberg's Arch Pharmacol 331(4):351–354

Kostyra E, Sienkiewicz-Szapka E, Jarmolowska B et al (2004) Opioid peptides derived from milk proteins. Pol J Food Nutr 13(54):25–35

Krasaekoopt W (2003) Yogurt from UHT milk: a review. Aust J Dairy Technol 58(1):26

Kunda PB, Benavente F, Catala-Clariana S et al (2012) Identification of bioactive peptides in a functional yogurt by micro liquid chromatography time-of-flight mass spectrometry assisted by retention time prediction. J Chromatogr A 1229:121–128

Lotfi B (2004) Optimization study for the production of an opioid-like preparation from bovine casein by mild acidic hydrolysis. Int Dairy J 14(6):535–539

Lottspeich F, Henschen A, Brantl V (1980) Novel opioid peptides derived from casein (beta-casomorphins). III. Synthetic peptides corresponding to components from bovine casein peptone. Hoppe Seylers Z Physiol Chem 361(12):1835–1839

McSweeney PLH (2004) Biochemistry of cheese ripening. Int J Dairy Technol 57(2–3):127–144

Meisel H, FitzGerald RJ (2000) Opioid peptides encrypted in intact milk protein sequences. Br J Nutr 84(1):27–31

Meltretter J, Schmidt A, Humeny A et al (2008) Analysis of the peptide profile of milk and its changes during thermal treatment and storage. J Agr Food Chem 56(9):2899–2906

Muehlenkamp MR, Warthesen JJ (1996) Beta-casomorphins: analysis in cheese and susceptibility to proteolytic enzymes from *Lactococcus lactis ssp Cremoris*. J Dairy Sci 79(1):20–26

Napoli A, Aiello D, Donna D et al (2007) Exploitation of endogenous protease activity in raw mastitic milk by MALDI-TOF/TOF. Anal Chem 79(15):5941–5948

Nguyen DD, Busetti F, Johnson SK et al (2018) Degradation of β-casomorphins and identification of degradation products during yoghurt processing using liquid chromatography coupled with high resolution mass spectrometry. Food Res Int 106:98–104

Nielsen MS, Martinussen T, Flambard B et al (2009) Peptide profiles and angiotensin-I-converting enzyme inhibitory activity of fermented milk products: effect of bacterial strain, fermentation pH, and storage time. Int Dairy J 19(3):155–165

Papadimitriou CG, Vafopoulou-Mastrojiannaki A, Silva SV et al (2007) Identification of peptides in traditional and probiotic sheep milk yoghurt with angiotensin I-converting enzyme (ACE)-inhibitory activity. Food Chem 105(2):647–656

Raies MH, Kapila R, Kapila S (2015) Release of β-casomorphin-7/5 during simulated gastrointestinal digestion of milk β-casein variants from Indian crossbred cattle (Karan fries). Food Chem 1 (168):70–79

Rizzello CG, Losito I, Gobbetti M et al (2005) Antibacterial activities of peptides from the water-soluble extracts of Italian cheese varieties. J Dairy Sci 88(7):2348–2360

Robinson RK (1995) A colour guide to cheese and fermented milks. Chapman and Hall, London

Sakurai T, Yamada A, Hashikura N et al (2018) Degradation of food-derived opioid peptides by bifidobacteria. Benef Microbes 9(4):675–682

Schieber A, Brückner H (2000) Characterization of oligo-and polypeptides isolated from yoghurt. Eur Food Res Technol 210(5):310–313

Sienkiewicz-Szłapkaa E, Jarmołowskaa B, Krawczuk S et al (2009) Contents of agonistic and antagonistic opioid peptides in different cheese varieties. Int Dairy J 19(4):258–263

Smacchi E, Gobbetti M (1998) Peptides from several Italian cheeses inhibitory to proteolytic enzymes of lactic acid bacteria, pseudomonas fluorescens ATCC 948 and to the angiotensin I-converting enzyme. Enzym Microb Technol 22(8):687–694

Stepaniak L, Fox PF, Sorhaug T et al (1995) Effect of peptides from the sequence 58-72 of beta-casein on the activity of endopeptidase, aminopeptidase, and X prolyl-dipeptidyl aminopeptidase from Lactococcus Lactis ssp. Lactis MG1363. J Agric Food Chem 43(3):849–853

Tamime AY, Robinson RK (1999) Yoghurt: science and technology. Woodhead Publishing Limited, Cambridge

Teschemacher H, Koch G (1991) Opioids in the milk. Endocrine Regul 25:147–150

Teschemacher H, Koch G, Brantl V (1997) Milk protein-derived opioid receptor ligands. Biopolymers 43(2):99–117

Toelstede S, Hofmann T (2008) Sensomics mapping and identification of the key bitter metabolites in Gouda cheese. J Agric Food Chem 56(8):2795–2804

Zioudrou C, Streaty RA, Klee WA (1979) Opioid peptides derived from food proteins. J Biol Chem 254(7):2446–2449

Biological Activities of Casomorphins 3

Abstract

β-Casomorphins (BCMs) demonstrate both protective and adverse effects through in vivo and in vitro trials. These peptides are significant in physiology, metabolism, immunomodulation, and inflammation. These peptides increase the induction of inflammatory response in mice gut through increased expression of inflammatory markers. They demonstrate a role in gut inflammation, gut motility, electrolyte absorption, lymphocyte stimulation and proliferation, histamine secretion, hormone release (gastrin, cholecystokinin, secretin, and gastrin inhibitory peptide), development of wheal and flare reactions, and mucus production. They are important for the regulation of fat intake and postprandial metabolism. These peptides show analgesic effects in addition to the manifestation of sedative activity. They demonstrate significant roles in sudden infant death syndrome (SIDS) and heart-related attributes. They produce cardiovascular activity and the creation of bradycardia and hypotension. BCM-7 and 10 established lymphocyte proliferation and stimulation at lower and higher doses, respectively. BCM-5 demonstrated both cardiodepressive and ionotropic effects at higher and lower doses, respectively.

3.1 Introduction

Casomorphins are opioid peptides derived from human and bovine milk. BCMs are the most significant group in casomorphins released from human and bovine β-casein milk protein. These exorphins are mostly released from β-casein of protein fraction and very few members have generated from the α-casein also. They are produced during gastrointestinal digestion, simulated gastrointestinal digestion, and fermentation. Regarding the bovine BCMs, the A1 milk hypothesis relates to the production of BCM-7 from only A1 β-casein. The latter protein shows histidine at position 67 to get easily cleaved to release this peptide. However, there is no any such hypothesis regarding the production of human BCMs. These peptides have been demonstrated to show morphine-like activity and show opioid activity. They possess all the structural features essential for the binding to the endogenous opioid receptors. These peptides are reported to show both beneficial and adverse effects. They have an essential role in physiological, metabolism, immunomodulation, and metabolism. These peptides increase the induction of inflammatory response in mice gut through increased expression of inflammatory markers like myeloperoxidase, antibodies, and toll-like receptors. They regulate fat intake and postprandial metabolism. They increase mucus production and secretion from the human and mouse cell lines. They demonstrate the role in the modulation of electrolyte absorption. Moreover, they establish analgesic effects in addition to the manifestation of sedative activity. These peptides have been linked to neurological manifestations like autism and schizophrenia. They produce cardiovascular activity and the creation of bradycardia and hypotension. Through the A1 milk hypothesis, BCM-7 has been correlated with heart diseases. On the other hand, BCM-5 demonstrated both cardiodepressive and ionotropic effects at higher and lower doses, respectively. BCM-7 and BCM-10 established lymphocyte proliferation and stimulation at lower and higher doses, respectively. BCM-7 increases histamine secretion in addition to the manifestation of wheal and flare reactions.

3.2 Gut Responses

3.2.1 Gut Inflammation

We have worked in the field of A1 and A2 cow milk and mice gut responses. We checked the inflammatory response of milk variants (A1 and A2 homozygous and A1/A2 heterozygous) in mice gut. The results were validated through the administration of chemically synthesized BCM peptides in the same mice models. The effect of milk variants and peptides was checked in vivo in mice models. The milk variants (A1A1, A1A2, and A2A2) were taken from already genotyped cows (Karan Frics). We selected gut for our study due to the reason that it is the most significant site for the bioavailability of the released peptides. We categorized the mice into four groups including three milk protein variants groups and a control group. The control group mice were fed with the basal diet and the experimental groups were fed basal diet replaced with these milk proteins. The immune response in the mice gut was checked through analysis of myeloperoxidase activity, monocyte chemotactic protein-1, cytokine profile, and immunoglobins (IgG, total IgE, IgG1, sIgA, and IgG2a). Moreover, the toll-like receptor (TLR-2 and TLR-4) mRNA expression was checked through qRT-PCR. The histopathology of the mice gut was performed for the goblet cells, IgA$^+$ cells, and total leukocytes. We found that oral administration of homozygous A1 and heterozygous A1A2 β-casein variants increased the production of myeloperoxidase, monocyte chemotactic protein-1, and interleukin-4. Besides, an increased antibody profile (IgG, IgG1, IgG2a, and total IgE) was also reported. Moreover, the histological observations showed increased infiltration of leucocytes in the mice gut. Furthermore, the consumption of A1 β-casein variants increased the expression of toll-like receptors (TLR-2 and TLR-4). Nevertheless, we could not observe any changes in IgA$^+$ cells, goblet cells, and sIgA on the consumption of A1 β-casein variants. Therefore, we concluded that consumption of A1A1 and A1A2 β-casein variants induced gut inflammatory response in mice through the Th2 pathway (Raies et al. 2014a).

The data generated in mice on the consumption of β-casein variants were validated through oral administration of chemically synthesized BCM-5 and BCM-7. It is assumed that these peptides are released from the A1 milk and may be responsible for the induction of these inflammatory responses. Therefore, the inflammatory response in mice's gut was checked on the administration of these peptides. The peptides are orally intubated to mice (7.5 × 10^{-8} mol/day/animal) for 15 days. The results of the study showed that oral intubation of these peptides increased the expression of inflammatory markers like myeloperoxidase, monocyte chemotactic protein-1, interleukin-4, and histamine. Moreover, the oral intubation of these peptides demonstrated an increased antibody profile including IgG and subtypes (IgG1/IgG2a) and total IgE. The histological examination revealed the leucocyte infiltration in mice gut. Nevertheless, there are no changes in sIgA, IgA$^+$ cells, and goblet cells. Therefore, these observations indicate that oral intubation of these peptides induced inflammation in mice through the Th2 pathway. Hence, the oral intubation of these peptides confirmed the results of the β-casein milk protein consumption (Raies et al. 2014b).

3.2.2 Gut Motility

BCM-7 decreased the gastrointestinal motility and the activity resembles the morphine (Shah 2000). These observations were confirmed in human subjects by Andiran and coworkers in 2003. When the human infants were fed with infant formula (cow milk-based), they observed constipation and showed anal fistulas. These effects were not observed when the infants were fed with mother's milk. Therefore, it is assumed that this effect occurs through the production of BCMs that bind the endogenous opioid receptor system (Andiran et al. 2003). Casein and whey protein suspensions were labeled with radioactive non-absorbable marker ^{141}Ce and intubated to rats through gastric route. The physiological parameters like gastrointestinal transit time (GITT) and Gastric emptying rate (GER) were lower on feeding whey compared to casein protein suspension. These biological parameters were completely (GITT) or partially (GER) diminished on pretreating animals (rats) with an opioid antagonist, naloxone. On administration of chemically synthesized BCM-4, non-significant changes were observed in these physiological parameters, however, substituted BCMs (BCM-4 and 5) showed significant changes while slowing down the GITT. These authors concluded that exorphins (BCMs) are released on gastrointestinal digestion of caseins that bind endogenous opioid receptors in the gut and suppress the gastrointestinal motility. The substituted exorphins are resistant to proteolysis and further demonstrate greater opioid activity

(Daniel et al. 1990). Special consideration was given to the genetic variants of β-casein, gastrointestinal time, brush border enzymes (dipeptidyl peptidase-4), and inflammatory response in rats. The observation revealed significant inhibition in the gastric transit and the increased expression of dipeptidyl peptidase-4 in the gut on feeding A1 milk. Moreover, increased expression of inflammatory marker and myeloperoxidase activity occurred on A1 milk consumption (Barnett et al. 2014). Moreover, reports have established that the consumption of (A1+A2) β-casein compared to A2 milk increased post-dairy digestive discomfort symptoms (PD3). The inflammatory markers increased in addition to the gastrointestinal transit times. Additionally, a decrease in the levels of short-chain fatty acids (SCFA) was observed. The consumption of (A1+A2) milk was associated with deterioration of PD3 symptoms in comparison with baseline in lactose tolerant and intolerant subjects. On the other hand, the consumption of A2 milk was unable to aggravate the PD3 symptoms. Overall, the investigators established that consumption of A1 milk was associated with increased inflammatory response, delay in gastrointestinal transit, deterioration of PD3 symptoms, and decreased cognitive dispensation quickness and accuracy (Jianqin et al. 2016).

3.2.3 Miscellaneous Gut Functions

Researchers have established that the peptides generated in the gut demonstrate gastrointestinal function before they are completely broken down to the individual amino acids and absorbed in the gut. These functions include the role of digestion and absorption in the gastrointestinal tract (Shimizu 2004). These peptides have a role in the regulation of the gastrointestinal system, secretion of proteins like mucins, and in turn formation of mucus. They may control the expression of hormones like gastrin, cholecystokinin, secretin, and gastrin inhibitory peptide (GIP). These may further help in the modification of gastric and pancreatic system secretion and ultimately regulate gastrointestinal motility. These gastrointestinal regulations may be manifested either through the individual amino acids or short-chain peptides that are potential ligands for the endogenous receptors system in the gut (Daniel et al. 1990). There are reports available that show casein stimulates the production and secretion of mucus in the gut. Therefore, it is assumed that this secretion is possible through the production of bioactive peptides and that in turn produces these physiological effects (Daniel et al. 1990; Yvon et al. 1994; Mahe et al. 1996). Caseinomacropeptide produced from the κ-casein through gastrointestinal enzymes is already reported to induce this biological effect (Yvon et al. 1994; Beucher et al. 1994; Pedersen et al. 2000). Similarly, the casocidin I (165–203) bioactive peptides generated from αS2-casein through chymosin proteolytic enzymes inhibit the growth of *Escherichia coli* and *Staphylococcus carnosus* (Zucht et al. 1995).

During the last few decades, more emphasis is given to the BCM peptides released from the β-casein during gastrointestinal digestion of milk. These peptides gained significant consideration due to the correlations between the release of these peptides and the increased risk of some human illnesses. These peptides resemble morphine in structure and demonstrate an affinity towards the endogenous opioid receptors. These opioid receptors are located in the gastrointestinal tract, central and peripheral nervous system, and some immune cells also. Therefore, these exorphins may regulate the gastrointestinal functions through the endogenous opioid receptor system (Baldi et al. 2005). BCMs modulate amino acid transport and expression of mucin proteins and formation of mucus in the gut (Brandsch et al. 1994; Claustre et al. 2002; Trompette et al. 2003; Zoghbi et al. 2006). The oral intubation of these peptides regulated the postprandial metabolism through somatostatin, pancreatic polypeptide, and insulin (Morley et al. 1988; Schusdziara 1983a, b; Takahashi et al. 1997; Froetschel 1996). These exorphins also suppressed the fat intake through the involvement of enterostatin (White et al. 2000). These peptides modulate electrolytes and water balance and antidiarrheal activity (Daniel et al. 1990). Moreover, these exorphins delay gastric emptying and in turn prolonged the gastrointestinal transit time (Daniel et al. 1990, 1990; Becker et al. 1990; Defilippi et al. 1995; Mihatsch et al. 2005; Shook and Burks 1986). The role of these peptides in electrolyte transport was assessed when chemically synthesized BCM-5 was applied from the mucosal side of the rabbit intestine. The results indicated decreased short-circuits and this effect reversed the application of naloxone (opioid antagonist). Therefore, it shows that this exorphin may modulate electrolyte transport in the gut (Hautefeuille et al. 1986). Schusdziara and co-investigators reported the role of BCMs in the stimulation of hormone release (insulin, somatostatin, and pancreatic polypeptide) in dogs (Schusdziara 1983a, b). The intraperitoneal or ICV route administration of these exorphins (BCM-5 and 7) stimulated the intake of fatty diet in satiated rat models (Lin et al. 1998). Moreover, BCM-7 was found to reverse the inhibitory activity of enterostatin on high-fat diet intake in rats (Christy et al. 2000). These exorphins increase the fat deposition in broiler chickens through the involvement of lipid metabolism genes (Chang et al. 2019). The correlation between consumption or release of these peptides in gut and regulation of electrolyte absorption is possibly a determining factor leading to antidiarrheal activity through endogenous opioid receptor systems (Tome et al. 1987; Mansour et al. 1988; Mahe et al. 1989).

3.3 Sudden Infant Death Syndrome (SIDS) and Heart-Related Attributes

The intracerebral-ventricle (ICV) administration of bovine BCMs (5 and 7) and morphiceptin to the rabbits (newborn) established a decrease in tidal volume and respiratory frequency. These respiratory attributes were dose-dependent, wherein BCM-5 was found 10 times more active compared to morphine. Nevertheless, the morphiceptin and BCM-7 demonstrated potency quite comparable to the morphine (Hedner and Hedner 1987). The direct administration of these exorphins (BCM-5 and 7) into the bloodstream showed a respiratory depression with morphiceptin and non-significant changes with the other two BCMs (BCM-5 and 7). These respiratory properties created with the food-derived exorphins like BCMs were reversed on opioid-antagonist (naloxone) administration. This phenomenon is further linked with the sudden infant death syndrome (SIDS). SIDS manifestations result in infants (healthy) that are below one year of age. These mortalities are not investigated completely even after autopsy analysis and assessment of clinical history (Sun et al. 2003). It is assumed that food-derived exorphins like BCMs may probably be the culprits in the stimulation of SIDS. The creation of a barrier between eminence and caudate nucleus was observed in experimental animals (rats) during the first postnatal weeks. In this perspective, they suggested that immature brain development may provide more chances for these exorphins to cross the BBB and enter the central nervous system (Peruzzo et al. 2000). Other studies have reported that there are no convincing observations that related increased opioid activity with SIDS manifestations. Nevertheless, the closeness between and near-miss SIDS and hyper endorphin syndrome leads to assumptions that enhanced opioid activity could be one of the risk factors for SIDS (Kuich and Zimmerman 1981). Overall, these observations establish that enhanced BCM opioid activity may further prompt the endogenous opioids in weakening breathing in the neonates.

Morphiceptin was found to cause bradycardia and hypotension in rats when this peptide was directly administered in the cerebroventricles or veins (Widy-Tyszkiewicz et al. 1986; Wei et al. 1980). On administration of the same peptide in the ventricles of hypertensive rats, dose-dependent increased blood pressure and heart rate were observed. In vitro studies in the guinea pigs, heart preparations on incubation with chemically synthesized BCM-5 revealed bipolar effect. The positive ionotropic activity and cardiodepressive response were observed at lower and higher doses, respectively (Liebmann et al. 1986).

3.4 Mucus Formation

Mucus is a jelly substance composed of diverse mucin proteins and other compounds. Mucin proteins are expressed and secreted by the goblet cells, predominantly found in the epithelial covering of the respiratory and intestinal systems. Mucus is considered as a protective covering against the pathogens in clearing off these microorganisms from the body. Therefore, this substance helps in the development of innate immunity. Trompette and coworkers in 2003 reported the secretion of mucins from the rat jejunum. The researchers checked the phenotypic expression of mucins through the endogenous opioid receptor system of the mice gut (Trompette et al. 2003). Three years later, other investigators performed in vitro experiments in rat and human goblet cell lines regarding mucin secretion with chemically synthesized BCM-7. They reported the expression of rMuc2 and rMuc3 from the rat DHE cell line with BCM-7. Moreover, increased expression of MUC5AC resulted from the human cell line HT29-MTX with this peptide (Zoghbi et al. 2006). As aforesaid the formation of mucus is beneficial or protective for animal and human health. In this perspective, animal deficient in the MUC2 gene is more exposed to the pathogenic bacteria and other risk factors. Therefore, there is an increased probability for these animals for the development of cancer and inflammation (Johansson et al. 2008). Therefore, it may be put forward that the milk-derived BCM-7 increases the production of mucins and in turn formation of mucus and ultimately helps the development of innate immunity (Trompette et al. 2003). These findings were further supported by the hypothesis laid by Bartley and McGlashan in 2010. These authors assumed that the milk-derived opioid peptides like BCM-7 may bind the endogenous opioid receptors and increase the production of mucins in the human subjects with leaky gut (Bartley and McGlashan 2010). In conclusion, this research area is still controversial and both the beneficial and adverse effects of these peptides are documented. Therefore, more research in animals and humans is needed to fully explore the area. Deeper research with the explored signal cascade at the cellular and molecular levels is needed to explore the mechanism and reach the conclusions.

3.5 Immunological Role

Reports have established the immunological perspective of cow milk-derived opioid peptides, more specifically BCM-7. These significant roles include the development of allergic and inflammatory responses, mucus formation, lymphocyte proliferation, histamine secretion, and skin reactions. The inflammatory response and mucus formation and its association with BCM-7 are discussed in the above sections. In this section, the other immunological attributes of the BCMs are discussed, comprehensively. From the immunological point of view, the most determining antigen that could evoke strong immune responses is the protein or protein-derived peptides. The most significant immune cells in the human body are the lymphocytes. When an antigen encounters a lymphocyte, the latter proliferates and this process is known as lymphocyte proliferation. The main objective of this proliferation is to increase the population of such cells to remove the antigens. BCMs were incubated with lamina propria-derived lymphocytes in vitro and suppression in the lymphocyte proliferation was reported with these peptides at low doses (10 mM). These results were validated while incubating these cells with naloxone (opioid antagonists) and that reversed the suppression of lymphocyte proliferation (Elitsur and Luk 1991). Human peripheral blood-derived lymphocytes were incubated with milk-derived exorphins (BCM-7 and BCM-10); lymphocyte stimulation and proliferation were observed at higher and lower doses, respectively (Meisel and Bockelmann 1999). Additionally, earlier reports have already detected the presence of these milk-derived exorphins (BCMs) in human intestinal aspirates (Svedberg et al. 1985). It is well-established that opioid peptides derived from milk have enough structural properties that could bind the endogenous opioid receptor system of immune cells and demonstrate lymphocyte stimulation or proliferation. Moreover, these opioid peptides are proline-rich and protease-resistant and demonstrate dual immunological properties including lymphocyte stimulation and proliferation (Migliore-Samour and Jolles 1988). The other researchers supported the findings while assuming these proline-rich milk protein-derived exorphins to demonstrate functionalities in cellular immunity (Zimecki et al. 1982). Nevertheless, more research with validated results is needed in this perspective especially the dosage that evokes such responses.

The milk-derived opioid peptides show many properties including small chain length (4–20), proline-rich and protease-resistance, and more especially their ability to bind endogenous opioid receptors on immune cells. These aforesaid attributes may enable these food-derived exorphins to be absorbed more rapidly in the gut (especially during leaky gut conditions), reach the blood, and bind immune cells and show numerous immunological responses. These responses may evoke both humoral and cell-mediated immune responses. The humoral immune response may stimulate B-cells to secrete antibodies and the latter may bind the mast cells to release histamine.

Earlier reports have already established the role of morphine in histamine production from mast cells in animal models and in vitro conditions. These allergic responses were further associated with anaphylactic reactions. BCM-7, the milk-derived agonist of morphine, when treated in vitro with human blood-derived peripheral leukocytes demonstrated a dose-dependent histamine release from these cells. These results were validated with human trials that showed the expression of wheal and flare reactions in the skin quite similar to histamine and codeine. Pre-treatment of the human subjects with terfenadine or cetirizine inhibited such immune responses. Moreover, the opioid antagonist (naloxone) also inhibited the release of histamine and the development of skin allergic reactions (Kurek et al. 1992). These results were further in agreement with the findings of Kurek and Malaczynska (1999) who observed a histamine secretion from the rat-derived peritoneal cells in vitro. Moreover, the treatment of sensitized guinea pigs with this peptide resulted in the wheal formation and an obstruction in the bronchia (Kurek and Malaczynska 1999). Subsequently, investigators established an association between serum DPPIV activities and BCM levels in mother's milk. They observed a decreased serum DPPIV activity in allergic infants whose mothers had higher BCM levels in milk. Additionally, the decreased content of BCMs (5 and 7) in mother's milk for allergic infants explains the absorption and transport of these exorphins to blood, decreased DPPIV activities, and enhanced BCM half-lives (Wasilewska et al. 2011a, b). These findings are in agreement with the data generated in our lab regarding BCM-5 and histamine release in vitro in cell lines. The results indicated a significant release of histamine and production of mast cell-specific tryptase from bone marrow-derived cells with BCM-5 (Reddi et al. 1983).

3.6   Miscellaneous Activities

BCMs show structural features that enable them to bind the endogenous opioid receptor system. These opioid receptors are distributed in the central and peripheral nervous system, gastrointestinal tract, and some immune cells. More recently in 2019 researchers have found that BCM-7 weakened sepsis-induced acute kidney injury (AKI) via reduction of the inflammatory response and oxidative stress. This effect was observed through the nuclear factor (NF-kb) pathway. Therefore, this report suggests a novel therapeutic potential of BCM-7 for lessening sepsis-induced kidney injury (Zhang et al. 2019). Earlier, reports have demonstrated the protective effect of BCM-7 in the inhibition of high glucose-induced renal proximal tubular epithelial-mesenchymal transition (EMT) partly through modulation of AngII-TGF-β1 pathway; however, this effect was not related to opioid receptor activation (Zhang et al. 2015). Moreover, Zhu and coworkers have demonstrated the protective effect of BCM-7 IN human lens epithelial cells (HLECs) against oxidative stress through increased expression of Foxo1 and in turn promoting Foxo1 nuclear translocation.

On the direct administration of these opioid peptides to blood, they showed analgesic and sedative activities (Paroli 1988). Among BCMs, BCM-7 demonstrated the prolonged effect for about 90 min or more; however, other BCMs (4–6) showed this effect for only 45 min (Matthies et al. 1984). Moreover, the administration of morphiceptin to rat neonates depicted hyperalgesic activity for four and a half months (Zadina et al. 1987). These exorphins, when consumed repeatedly, may show physical dependence and tolerance. The tolerance was reported when morphiceptin was administered repeatedly. The doses (0.5 µl) were administered intrathecally for 3.5–4.5 days in rats. An association was established between morphiceptin and physical dependence when administered (1–23 µl) per hour for 70–72 h in rats. This exorphin was injected at the junction between the fourth ventricle and aqueduct in rats. The confirmation was done with the administration of naloxone (4 mg/kg). Therefore, the authors concluded that prolonged consumption of these exorphins may induce physical dependence like other opioids (Chang et al. 1983).

References

Andiran F, Dayi S, Mete E (2003) Cow Milk consumption in constipation and anal fissures in infants and young children. J Paediatr Child Health 39:329–331

Baldi A, Ioannis P, Chiara P et al (2005) Biological effects of milk proteins and their peptides with emphasis on those related to the gastrointestinal ecosystem. J Dairy Res 72:66–72

Barnett MP, McNabb WC, Roy NC et al (2014) Dietary A1 β-casein affects gastrointestinal transit time, dipeptidyl peptidase-4 activity, and inflammatory status relative to A2 β-casein in Wistar rats. Int J Food Sci Nutr 65:720–727

Bartley J, McGlashan SR (2010) Does milk increase mucus production? Med Hypotheses 74:732–734

Becker A, Hempel G, Grecksch G et al (1990) Effects of beta-casomorphin derivatives on gastrointestinal transit in mice. Biomed Biochim Acta 49(11):12031–12037

Beucher S, Levenez F, Yvon M et al (1994) Effects of gastric digestive products from casein on CCK release by intestinal cells in the rat. J Nutr Biochem 5(12):578–584

Brandsch M, Brust P, Neubert K et al (1994) β-Casomorphins–chemical signals of intestinal transport systems. In: Beta-casomorphins and related peptides: recent development. Wiley, Weinheim, p 209

Chang KJ, Wei ET, Killian A et al (1983) Potent morphiceptin analogs: structure-activity relationships and morphine-like activities. J Pharmacol Exp Ther 227:403–408

Chang WH, Zheng AJ, Chen ZM et al (2019) β-Casomorphin increases fat deposition in broiler chickens by modulating the expression of lipid metabolism genes. Animal 13(4):777–783. https://doi.org/10.1017/S1751731118002197

Christy L, White GA, Bray A et al (2000) Intragastric β-casomorphin-7 attenuates the suppression of fat intake by enterostatin. Peptides 21:1377–1381

Claustre J, Toumi F, Trompette A et al (2002) Effects of peptides derived from dietary proteins on mucus secretion in rat jejunum. Am J Physiol Gastrointest Liver Physiol 283(3):521–528

Daniel H, Vohwinkel M, Rehner G (1990) Effect of casein and beta-casomorphins on gastrointestinal motility in rats. J Nutr 120(2):252–257

Defilippi C, Gomez E, Charlin V et al (1995) Inhibition of small intestinal motility by casein: a role of β-casomorphins? Nutrition 11(6):751–754

Elitsur Y, Luk GD (1991) Beta-casomorphin (BCM) and human colonic lamina propria lymphocyte proliferation. Clin Exp Immunol 85:493–497

Froetschel MA (1996) Bioactive peptides in digests that regulate gastrointestinal function and intake. Am Soc Anim Sci 74(10):2500–2508

Hautefeuille M, Brantl V, Dumontier AM et al (1986) In vitro effects β-casomorphins on ion transport in rabbit ileum. Am J Phys 250:G92–G97

Hedner J, Hedner T (1987) Beta-casomorphins induce apnea and irregular breathing in adult rats and newborn rabbits. Life Sci 41(20):2303–2312

Jianqin S, Leiming X, Lu X et al (2016) Effects of milk containing only A2 beta-casein versus milk containing both A1 and A2 beta-casein proteins on gastrointestinal physiology, symptoms of discomfort, and cognitive behavior of people with self-reported intolerance to traditional cows' milk. Nutr J 15(1):45

Johansson MEV, Phillipson J, Petersson A et al (2008) The inner of the two Muc2 mucin-dependent mucus layers in colon is devoid of bacteria. Proc Acad Nat Sci 105:15064–15069

Kuich TE, Zimmerman C (1981) Could endorphins be implicated in sudden infant death syndrome? New Eng Med 304:973–976

Kurek M, Malaczynska T (1999) Food allergy, atopic dermatitis-nutritive casein formula elicits pseudoallergic skin reactions by prick testing. Int Arch Allergy Immunol 118(2):228–229

Kurek M, Przybilla B, Hermann K et al (1992) A naturally occurring opioid peptide from cow's milk, beta-casomorphine-7, is a direct histamine releaser in man. Int Arch Allergy Immunol 97:115–120

Liebmann C, Barth A, Neubert K (1986) Effects of β-casomorphin on 3H-ouabain binding to Guinea-Pig heart membranes. Pharmazie 41:670–671

Lin L, Umahara M, York DA et al (1998) β-Casomorphins stimulate and enterostatin inhibits the intake of dietary fat in rats. Peptides 19:325–331

Mahe S, Tomé D, Dumontier AM et al (1989) Absorption of intact beta-casomorphins (beta-CM) in rabbit ileum in vitro. Reprod Nutr Dev 29(6):725–733

Mahe S, Roos N, Benamouzig R et al (1996) Gastrojejunal kinetics and the digestion of [N-15] beta-lactoglobulin and casein in humans: the influence of the nature and quantity of the protein. Am J Clin Nutr 63(4):546–552

Mansour A, Tome D, Rautureau M et al (1988) Luminal anti-secretory effects of a [beta]-casomorphin analog on rabbit ileum treated with cholera toxin. Pediatr Res 24(6):751

Matthies H, Stark H, Hartrodt B (1984) Derivatives of β-casomorphins with high analgesic potency. Peptides 5:463–470

Meisel H, Bockelmann W (1999) Bioactive peptides encrypted in milk proteins: proteolytic activation and thropho-functional properties. Antonie Van Leeuwenhoek 76(1):207–215

Migliore-Samour D, Jolles P (1988) Casein, a prohormone with an immunomodulating role for the newborn? Exp Dermatol 44:188–193

Mihatsch WA, Franz AR, Kuhnt B et al (2005) Hydrolysis of casein accelerates gastrointestinal transit via reduction of opioid receptor agonists released from casein in rats. Biol Neonate 87(3):160–163

Morley JE, Levine AS, Yamada et al (1988) Effect of exorphins on gastrointestinal function, hormonal release, and appetite. Gastroenterology 84(6):1517–1523

Paroli E (1988) Opioid peptides from food (exorphins). World Rev Nutr Diet 55:58–63

Pedersen NL, Nagain-Domaine C, Mahe S et al (2000) Caseinomacropeptide specifically stimulates exocrine pancreatic secretion in the anesthetized rat. Peptides 21(10):1527–1535

Peruzzo B, Pastor FE, Blazquez JL et al (2000) A second look at the barriers of the medial basal hypothalamus. Exp Brain Res 132:10–26

Raies MH, Kapila R, Sharma R et al (2014a) Comparative evaluation of cow β-casein variants (a1/a2) consumption on th2-mediated inflammatory response in mouse gut. Eur J Nutr 53(4): 1039–1049

Raies MH, Kapila R, Saliganti V (2014b) Consumption of β-casomorphins-7/5 induce inflammatory immune response in mice gut through th2 pathway. J Funct Foods 8:150–160

Reddi S, Kapila R, Dang AK et al (1983) Evaluation of the allergenic response of milk bioactive peptides using mouse mast cell. Milchwissenschaft 67(2):117–232

Schusdziara V (1983a) Effect of beta-casomorphins on somatostatin release in dogs. Endocrinology 112(6):1948–1951

Schusdziara V (1983b) Effect of beta-casomorphins and analogs on insulin release in dogs. Endocrinology 112(3):885–889

Shah N (2000) Effects of milk-derived bioactives: an overview. Br J Nutr 84:S3–S310

Shimizu M (2004) Food-derived peptides and intestinal functions. Biofactors 21(1–4):43–47

Shook JE, Burks TF (1986) Peripheral mu-opioid receptors mediate small intestinal transit (sit) but not analgesia in mice. Fed Proc 45:1344–1351

Sun Z, Zhang Z, Ang X et al (2003) Relation of beta-casomorphin to apnea in sudden infant death syndrome. Peptides 24(6):937–943

Svedberg J, de Haas J, Liemenstoff G et al (1985) Demonstration of beta-casomorphin immunoreactive materials in vitro digests of bovine milk and in small intestine contents after bovine milk ingestion in adult humans. Peptides 6:825–830

Takahashi M, Moriguchi S, Suganuma H et al (1997) Identification of casoxin C, an ileum-contracting peptide derived from bovine kappa-casein, as an agonist for C3a receptors. Peptides 18(3):329–336

Tome D, Dumontier AM, Hautefeuille M et al (1987) Opiate activity and transepithelial passage of intact beta-casomorphins in rabbit ileum. Am J Physiol Gastrointest Liver Physiol 253 (6):737–744

Trompette A, Claustre J, Caillon F et al (2003) Milk bioactive peptides and β-casomorphins induce mucus release in rat jejunum. J Nutr 133(11):3499–3503

Wasilewska J, Kaczmarski M, Kostyra E et al (2011a) Cow's-milk-induced infant apnoea with increased serum content of bovine β-casomorphin-5. J Pediatr Gastroenterol Nutr 52:772–775

Wasilewska J, Sienkiewicz SE, Kuzbida E et al (2011b) The exogenous opioid peptides and DPP IV serum activity in infants with apnoea expressed as apparent life-threatening events (ALTE). Neuropeptides 43(3):189–195

Wei ET, Lee A, Chang JK (1980) Cardiovascular effects of peptides related to the enkephalins and β-casomorphin. Life Sci 26:1517–1522

White CL, Bray GA, York DA (2000) Intragastric beta-casomorphin (1-7) attenuates the suppression of fat intake by enterostatin. Peptides 21(9):1377–1381

Widy-Tyszkiewicz E, Czonkowski A, Szreniawski Z (1986) Cardiovascular response to morphiceptin in spontaneously hypertensive and normotensive rats. Pol J Pharmacol Pharm 38:51–56

Yvon M, Beucher S, Guilloteau P et al (1994) Effects of caseinomacropeptide (CMP) on digestion regulation. Reprod Nutr Dev 34(6):527–537

Zadina JE, Kastin AJ, Manasco PK et al (1987) Long-term hyperalgesia induced by neonatal β-endorphin and morphiceptin is blocked by neonatal Tyr-MIF-1. Brain Res 409:10–18

Zhang W, Song S, Liu F et al (2015) Beta-casomorphin-7 prevents epithelial-mesenchymal trans differentiation of NRK-52E cells at high glucose level: Involvement of AngII-TGF-β1 pathway. Peptides 70:37–44

Zhang Z, Zhao H, Ge D, Wang S et al (2019) β-Casomorphin-7 ameliorates sepsis-induced acute kidney injury by targeting the NF-κB pathway. Med Sci Monit 25:121–127

Zimecki M, Janusz M, Staroscik K et al (1982) Effect of a proline-rich polypeptide on donor cells in graft-versus-host reaction. Immunology 47:141–145

Zoghbi S, Trompette A, Claustre J et al (2006) β-Casomorphin-7 regulates the secretion and expression of gastrointestinal mucins through a μ-opioid pathway. Am J Physiol Gastrointest Liver Physiol 290(6):1105–1113

Zucht HD, Raida M, Adermann K et al (1995) Casocidin-I: a casein-αs2 derived peptide exhibits antibacterial activity. FEBS Lett 372(2-3):185–188

Lactorphins and Lactoferroxins

4

Abstract

Lactorphins and lactoferroxins are the food exorphins released from the whey proteins of milk. α-Lactorphin and β-lactorphin are released from α-lactalbumin and β-lactoglobulin, respectively, of milk proteins. Similarly, lactoferroxins are released from iron-containing lactoferrin of milk protein. α-Lactorphin (Tyr-Gly-Leu-Phe-NH$_2$) and β-lactorphin (Tyr-Leu-Leu-Phe-NH$_2$) are amidated protein fragments from 69 to 72 and 118 to 121, respectively, from α-lactalbumin and β-lactoglobulin. These act as agonists for μ-opioid receptors. Similarly, lactoferroxin A (YLGSGY-OCH$_3$), B (RYYGY-OCH$_3$), and C (KYLGPQYOCH$_3$) are carboxy-methylated protein fragments from 118 to 121, 536 to 540, and 673 to 679, respectively. These peptides show ACE inhibitory activity with IC$_{50}$ 171.8, 733.3, and 1153.2 μM, respectively. α-Lactorphin has a favorable role in endothelial function and in turn cardiovascular functions. β-Lactorphin demonstrated osteoprotective potential in decreasing inflammatory status and increasing bone formation markers. Moreover, α-lactorphin has enough scope in up-regulating the expression of mucin gene and in turn maintenance of mucosal immunity.

4.1 Introduction

Milk is a source of proteins that contains all essential amino acids. Milk possesses two protein fractions, referred to as caseins and whey proteins. Both the fractions upon gastrointestinal hydrolysis release amino acids and consequently form an amino acid pool in humans. These amino acids are used for the synthesis of human protein inside our bodies. Besides nutritional role, these proteins produce peptides in the gut and that is biologically active. The active peptides help in regulating numerous physiological processes. They have antihypertensive, analgesic, mineral binding, immunomodulating, and other activities in the human body.

M. R. Ul Haq, *Opioid Food Peptides*, https://doi.org/10.1007/978-981-15-6102-3_4

Besides these functions, many food-derived peptides have the affinity to bind the opioid receptors. These include μ (MOP), δ (DOP), and κ (KOP), and different food exorphins bind with these receptors. These receptors are located in the central and peripheral nervous system, gastrointestinal tract, and some immune cells. The caseins comprise about 80% of total proteins in cow milk that is further divided into αs1, αs2, β, and γ subtypes. The whey proteins include about 20% and include proteins like α-lactalbumin, β-lactoglobulin, immunoglobulins, lactoferrin, lactoperoxidase, etc. Reports have established that peptides derived from the whey proteins also demonstrate opioid activity. The opioid peptide-like α-lactorphin (Tyr-Gly-Leu-Phe-NH$_2$) is derived from bovine milk protein α-lactalbumin with amidated carboxyl-terminal. Similarly, β-lactorphin (Tyr-Leu-Leu-Phe-NH$_2$) is released from bovine milk protein β-lactoglobulin. These milk-derived exorphins establish a low affinity towards opioid receptors demonstrated on guinea pig ileum (GPI) assay (Garg et al. 2016; Teschemacher et al. 1997). Besides these food exorphins, synthesized opioid peptides corresponding to human lactoferrin are also available. These milk-derived opioid peptides are known as lactoferroxin and there are three types of such peptides. Those are known as lactoferroxin A (Tyr-Leu-Gly-Ser-Gly-Tyr-OCH$_3$), B (Arg-Tyr-Tyr-Gly-Tyr-OCH$_3$), and C (Lys-Tyr-Leu-Gly-Pro-Gln-Tyr-OCH$_3$). These peptides have methoxylated carboxyl-terminal. These peptides are opioid antagonists like that of casoxins. These peptides act as moderate opioid antagonists (Garg et al. 2016). Literature has depicted that most of the milk-derived opioid peptides demonstrate μ (MOP) opioid receptor selectivity. However, the opioid peptides released from κ-casein that are known as casoxins establish affinity for both μ- and δ-opioid receptors when observed through MVD and GPI assays (Teschemacher et al. 1997).

4.2 Structure of Lactorphins and Lactoferroxins

4.2.1 Structure of Lactorphins

The whey protein α-lactalbumin from the cow milk yields opioid peptide α-lactorphin on digestion. The opioid peptide is composed of four amino acids, tyrosine at the amino-terminal, phenylalanine at the carboxyl-terminal and glycine and leucine in between these amino acids. The opioid peptide is amidated at the carboxyl-terminal. It has a sequence (Tyr-Gly-Leu-Phe-NH$_2$), molecular formula $C_{26}H_{35}N_5O_5$, and molecular weight 497.6 g/mol. It is derived from the α-lactalbumin with the fragment 69–72. It acts as an agonist and has selectivity for the μ-opioid receptor (Teschemacher et al. 1997; Day et al. 1981).

Similarly, the cow milk β-lactoglobulin on gastrointestinal digestion yields opioid peptide β-lactorphin. It has a sequence (Tyr-Leu-Leu-Phe-NH$_2$). This peptide is composed of four amino acids, tyrosine at the amino-terminal, phenylalanine at the carboxyl-terminal and two leucine residues in between these amino acids. The opioid peptide is amidated at the carboxyl-terminal. It has a molecular formula $C_{30}H_{43}N_5O_5$ and a molecular weight of 553.7 g/mol. It is derived from the β-lactoglobulin with the fragment 118–121. It acts as an agonist and has selectivity for the μ-opioid receptor (Teschemacher et al. 1997; Day et al. 1981). The structure of these two opioid peptides is given in Fig. 4.1.

4.2.2 Structure of Lactoferroxins

Lactoferroxins are milk-derived opioid peptides from lactoferrin. The latter is a globular secretary protein found in various fluids including milk, tears, saliva, and nasal secretions. Three types of exorphins are derived from this protein in milk known as lactoferroxin A, B, and C. All the three forms are carboxy-methylated at the carboxyl-terminal. Type A is a lactoferrin fragment from 318 to 325 composed of six amino acid residues with a sequence (YLGSGY-OCH$_3$). It has a molecular formula $C_{32}H_{44}N_6O_{10}$ and molecular weight 672.7 g/mol. Type B is a lactoferrin fragment from 536 to 540 composed of five amino acid residues with a sequence (RYYGY-OCH$_3$). It has a molecular formula $C_{36}H_{46}N_8O_9$ and molecular weight 734.8 g/mol. Type C is a lactoferrin fragment from 673 to 679 composed of seven amino acid residues with a sequence (KYLGPQYOCH$_3$). It has a molecular formula $C_{43}H_{63}N_9O_{11}$ and molecular weight 882 g/mol. The peculiarities of all the three opioid peptides are their antagonistic characteristic feature and have μ-opioid receptor selectivity (Tani et al. 1990). The structure of these opioid peptides is given in Fig. 4.1.

4.3 Classification of Lactorphins and Lactoferroxins

Lactorphins are classified into two types and generally regarded as alpha (α) and beta (β) forms. As mentioned in the previous section, alpha (α)-lactorphin is a four amino acid peptide fragment of α-lactalbumin between 69 and 72 and amidated at the carboxyl-terminal. Similarly, the beta (β)-lactorphin is also a four amino acid peptide derived from β-lactoglobulin between 118 and 121 and amidated at the carboxyl-terminal. Both types of lactorphins are agonists and have an affinity for the μ-opioid receptor (Teschemacher et al. 1997; Day et al. 1981).

Lactoferroxins are classified into three forms as A, B, and C. All the three forms A, B, and C are derived from the lactoferrin of milk protein and represent fragments 318–323, 536–540, and 673–679, respectively. However, compared to other milk peptides these act as the antagonist and have an affinity for μ-opioid receptors (Tani et al. 1990). The classification is given in Table 4.1.

4.4 Opioid Activity of Lactorphins and Lactoferroxins

The production, properties, and biological functions of food-derived exorphins have been subject to research for the last almost 40 years. Many of these peptides from milk have selectivity towards the endogenous opioid receptor system of humans.

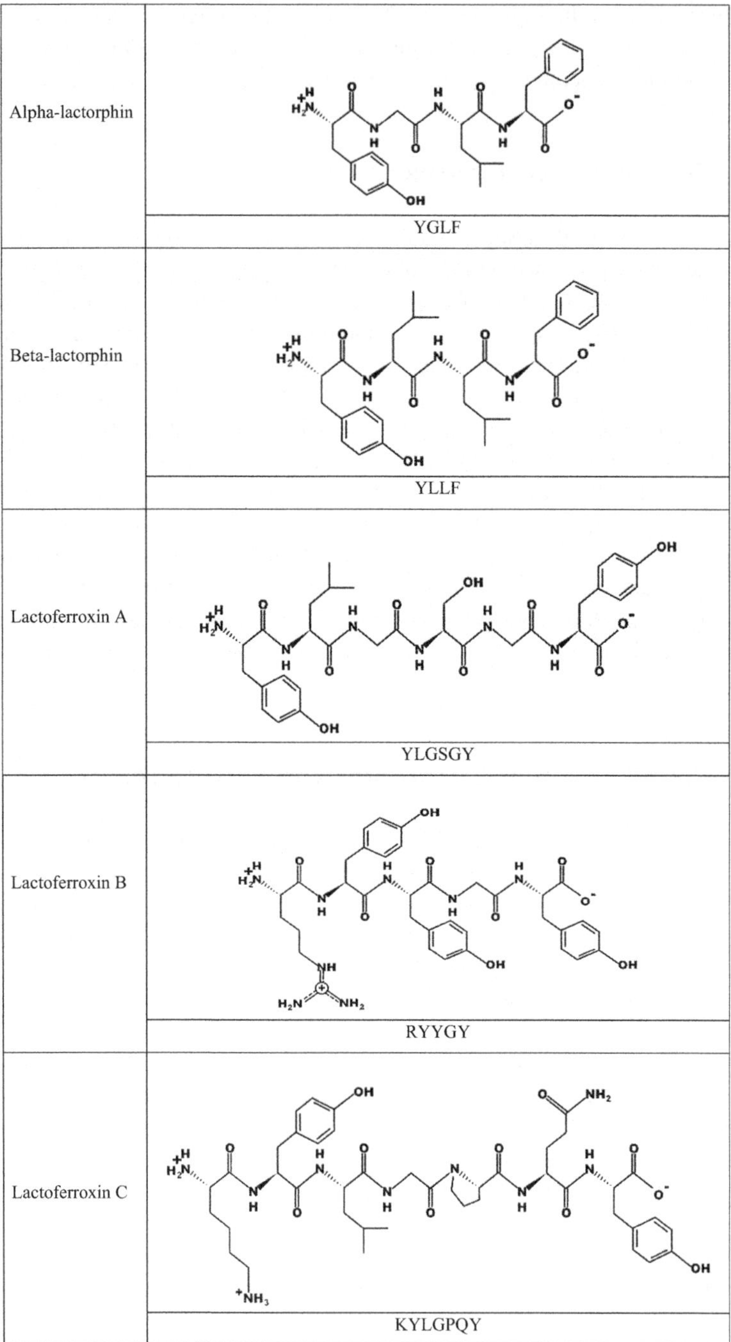

Fig. 4.1 Structures of lactorphins and lactoferroxins

Table 4.1 Classification of lactorphins and lactoferroxins

Exorphins	Protein precursor	Amino acid sequence	Opioid receptor selectivity	Agonist/antagonist	References
Lactorphins					
α-Lactorphin	α-Lactalbumin (69–72)	YGLF-NH$_2$	μ	A	Teschemacher et al. (1997)
β-Lactorphin	β-Lactoglobulin (118–121)	YLLF-NH$_2$	μ	A	Day et al. (1981)
Lactoferroxins					
Lactoferroxin A	Lactoferrin (318–323)	YLGSGY-OCH$_3$	μ	AN	Tani et al. (1990)
Lactoferroxin B	Lactoferrin (536–540)	RYYGY-OCH$_3$	μ	AN	
Lactoferroxin C	Lactoferrin (673–679)	KYLGPQY-OCH$_3$	μ	AN	

These peptides have demonstrated a role in the management of pain and behavioral processes. There is not any available peptide-based therapeutics for controlling the nervous system actions. Among many reasons available, one of the strong justifications is the hindrance of these peptides to cross the blood–brain barrier (BBB). The aforesaid barrier prevents these peptides to reach the central nervous system. If or when these food-derived exorphins cross this barrier bind the opioid receptors in the brain and regulate the nervous functions. Before that, they may exert their functions in the gastrointestinal tract, blood cells, and some immune cells. Among many advantages are the stability and lower side effects of food-derived opioid peptides in comparison to the synthetic and endogenous opioids. These peptides have a tyrosine amino acid residue at the N-terminal and an aromatic amino acid residue at the 3rd or 4th position (Teschemacher et al. 1997). These peptides have pharmacological properties similar to opium and act like morphine in the brain. Opioid peptides are derived from milk, wheat, soy, rice, spinach, and other food sources. Opioid peptide sequences are encrypted in the whey proteins like α-lactalbumin and β-lactoglobulin (Belem et al. 1999). These food-derived exorphins are biologically active and demonstrate various physiological functions. These are potent and act at micromolar quantities to demonstrate biological effects (Meisel and Fitzgerald 2000). The exorphins from the milk act agonist at μ, δ, and κ-opioid receptors. Although lactorphins act as agonists and bind μ-receptors, the lactoferroxins serve as antagonists with μ-opioid receptor selectivity.

4.5   Production of Lactorphins and Lactoferroxins

From a nutritional perspective, nowadays, the whey protein part of milk also known as plasma of milk parallels the other proteins like casein, soy, and egg albumin. Besides this, it has enough scope and potential for the regulation of physiological functions. It consists of α-lactalbumin (20%), β-lactoglobulin (50%), bovine serum albumin (12%), and immunoglobulins (8%) of the total whey proteins present in cow milk. The α-lactalbumin powerfully binds with calcium with a dissociation constant ranging from 10 to 12M and has an important role in the synthesis of lactose. The β-lactoglobulin is usually present in the form of a dimer (2βlgA/2βlgB or 1βlgA/β lgB). This protein has a significant role in the transport of vitamin A. The b-LG is normally present as a dimer, consists of two b-LGA, two b-LGB, or one b-LGA, and one b-LGB, and it may be involved in vitamin A transport. Whey also contains lactoferrin that stores iron and is a biologically active molecule.

Belem and coworkers in 1999 performed fermentation (fed-batch) using *Kluyveromyces marxianus* var. *marxianus,* for the formation of whey-soluble proteins (oligopeptides). The investigators used media containing whey (5–15%) for fermentation with the aforesaid bacteria for 2 h. This was followed by fed-batch fermentation for 50 h. The wort was harnessed for the analysis of oligopeptides with HPLC. While using this bioanalytical technique, four peptide fractions were collected at different retention times. The peptide fractions were analyzed by mass spectroscopy (API). The sequences of peptides from the whey proteins were compared with known physiologically active peptides. Two bioactive peptide fragments were detected as precursor sequences with inherent β-lactorphin (YLLF) sequences. These bioactive peptides further could have antihypertensive properties (Belem et al. 1999).

The hydrolysis of the goat whey was done by a proteolytic enzyme, pepsin in an ultrafiltration reactor (continuous). The hydrolysate was analyzed by reverse phase-high pressure liquid chromatography (RP-HPLC). This bioanalytical technique separates the peptides based on hydrophobic nature. The mass spectroscopy further showed the presence of physiologically active peptide α-lactorphin (Bordenave et al. 1999).

Tani and coworkers in 1991 tried to evaluate the presence of lactoferroxin A, B, and C from the pepsin digest of milk lactoferrin. In this perspective, the solution of lactoferrin (human) was made by the method of Bläckberg and Hernell (1980). The purity of the protein was checked with SDS-PAGE that revealed a single band at 80 kDa. A solution of lactoferrin was prepared by dissolving 80 mg of lactoferrin in 8 ml distilled water and pH maintained at 1.4 with hydrochloric acid (concentrated). The hydrolysis was performed with pepsin (1.6 mg) at 37 °C for 6 h. To inactivate the pepsin after complete incubation sodium hydroxide was added to increase the pH. This was followed by the boiling of the hydrolysate for 10 min and then lyophilized. The hydrolysate powder was treated with methanol for esterification with the addition of hydrogen chloride (70 mM) under anhydrous conditions. The insoluble material was removed by centrifugation and the supernatant was run on the octadecyl silica (ODS) column with a linear gradient (0–50%) of acetonitrile added with the 0.1% trifluoroacetic acid and the run continued for 50 min. The individual fractions were collected and dried on the vacuum evaporator. The eluted powdered fractions were subjected for sequence determination on the mass spectroscopy and opioid activity assessment with pharmacological technique. The results indicated an opioid activity in the peptic digests from human lactoferrin. Moreover, three bioactive peptides were separated on RP-HPLC, between 318 and 323, 536 and 540, and 673 and 679 and corresponded to the human lactoferrin protein. Mass spectroscopy revealed the sequences as Tyr-Leu-Gly-Ser-Gly-Tyr-OCH$_3$, Arg-Tyr-Tyr-Gly-Tyr-OCH$_2$, and Lys-Tyr-Leu-Gly-Pro-Gln-Tyr-OCH$_3$. These opioid peptides were named as lactoferroxin A, B, and C. The opioid activity was visualized with the membranes from rat brain. The antagonistic activity for the μ-receptors was checked on guinea pig ileum, for δ-receptors the activity was assessed in mouse vas deferens, and for κ-receptors was evaluated on rabbit vas deferens. The IC$_{50}$ values were 15, 10, and 23 mM for lactoferroxin A, B, and C, respectively. It was also found that lactoferroxin A has more affinity towards μ-receptors, while B and C have more selectivity towards the κ-receptors. Furthermore, the correlation between the structure and activity revealed that all three types of lactoferroxins may be represented by a general formula XA-Tyr-XB-Tyr-OCH$_3$. The amino acid residue at the position XA could affect the selectivity of the antagonist protein fragments for opioid receptor subtypes (Tani et al. 1990).

4.6   Physiological Effects of Lactorphins and Lactoferroxins

As aforementioned these bioactive peptides act as agonists (lactorphins) and antagonists (lactoferroxins) for the μ-receptors located in the central and peripheral nervous system, gastrointestinal tract, and some immune cells. Therefore, the binding between exorphins and their receptors leads to the manifestations of various physiological functions.

4.6.1   Cardiovascular Functions

Alpha-lactorphin (Tyr-Gly-Leu-Phe) and beta-lactorphin (Tyr-Leu-Leu-Phe) as aforesaid are tetrapeptides generated from the alpha-lactalbumin and beta-lactoglobulin, respectively, with enzymatic cleavage. These have been demonstrated to decrease the blood pressure in spontaneously hypertensive rats (SHR) (30–35 weeks old). Researchers have investigated the antihypertensive function of these peptides in rats with well-established hypertension and compared with similar aged Wistar-Kyoto (WKY) rats (normotensive) in vitro. The investigators made preparations of mesenteric arteries of both SHR and WKY rats. They observed a decrease in the relaxation to acetylcholine in SHR compared to WKY rats that were endothelium-dependent. It was reported that acetylcholine-induced relaxation in SHR was amplified with incubation of alpha-lactorphin or beta-lactorphin. Nitric acid (NO) that is formed by nitric acid synthase (NOS) has a significant role in smooth muscle relaxation. In this perspective, it was suggested that NO has a potential role in this regard as this effect was eliminated with the inhibitor of NOS (N(G)-nitro-L-arginine methyl ester). The acetylcholine-induced relaxation was not altered with the addition of tetra-ethyl ammonium (TEA), an inhibitor of potassium channel inhibitor. Moreover, the addition of diclofenac did not change the acetylcholine-induced relaxation. It was therefore clear that endothelial vasodilatory prostanoids do not supplement the food-derived tetrapeptide in the vascular relaxation. The endothelium-independent relaxations were augmented in the presence of sodium nitroprusside (NO donor) in the mesenteric arterial preparations derived from SHR by co-incubation with beta-lactorphin. However, these lactorphins (alpha and beta) could not change these endothelium-dependent relaxations in the mesenteric arterial preparations derived from WKY. Therefore, these observations establish that both these peptides have the potential to improve the vascular relaxation in adult spontaneously hypertensive rats in vitro. Alpha-lactorphin was engaged in the direction of endothelial function and beta-lactorphin increased the relaxation that was endothelium-independent (Sipola et al. 2002). Similarly, Nurminen and coworkers in 2000 studied the effect of the commercially synthesized peptide with a sequence similar to that of bovine milk-derived alpha-lactorphin (Tyr-Gly-Leu-Phe). The animal trials were performed in SHR and WKY rats that were conscious. The animals were under continuous radiotelemetric monitoring observations. The synthetic peptide decreased the blood pressure in a dose-dependent way; however, no changes were observed in the heart rate in both the models of rats. The lowest concentration of this peptide that decreased the blood pressure was found to be 10 µg/kg. On the other hand, a dose of 100 µg/kg was required to demonstrate the highest reductions in systolic and diastolic blood pressure (by 23 ± 4 and 17 ± 4 mmHg, respectively) in SHR. Furthermore, when a dose greater than 1 mg/kg was administered there were no further reductions observed. Alpha-lactorphin could not show significant differences in the reduction of BP between WKY and SHR. The opioid antagonist, naloxone eradicated the reductions in the BP induced by the opioid agonist, alpha-lactorphin. These observations lead to the conclusion that this peptide through the involvement of opioid receptors has a potential significance in the maintenance of cardiovascular function (Nurminen et al. 2000).

4.6.2 Miscellaneous Functions

Earlier studies have well-established that peptides derived from the alpha-lactalbumin and beta-lactoglobulin demonstrated the opioid activities. These studies were followed by other researchers who tried to establish other activities of these peptides as well. The commercially synthesized alpha-lactorphin (Tyr-Gly-Leu-Phe) and beta-lactorphin (Tyr-Leu-Leu-Phe) showed angiotensin-I-converting enzyme (ACE) inhibitory activities. The peptides alpha-lactorphin and beta-lactorphin showed ACE inhibitory activities with IC_{50} values 171.8, 733.3, and 1153.2 µM, respectively (Mullally et al. 1997).

The researchers have tried to explore the role of β-lactorphin in bone health. The potential role of this peptide was assessed in bone remodeling in the osteoporotic rat model. These animals were administered with this peptide at a dose of 50 µg/kg of mice per day for 8 weeks. The parameters like trabecular microarchitectural of femoral and tibiae bone were assessed through a micro-CT scan. Enzyme-linked immunosorbent assay (ELISA) was performed to check the turnover markers like alkaline phosphatase, receptor activator of nuclear factor kappa-B ligand (RANKL), and osteocalcin. Moreover, other parameters like bone formation and resorption were also studied. In addition to these inflammatory markers including TGF-β, TNF-α and IFN-γ were estimated by ELISA. Results indicated that this milk-derived peptide inhibited the increase in inflammatory cytokines and bone turnover. However, the parameters related to bone formation markers were elevated. The change in shape and size of tibiae and femoral bones due to ovariectomy lessened through β-lactorphin administration. Therefore, these observations demonstrate the osteoprotective potential of this peptide in decreasing inflammatory status and increasing bone formation markers (Pandey et al. 2018).

The hypothesis that milk protein-derived peptides induce mucus secretion through the expression of mucins was investigated in vitro in a human colonic epithelium cell line (HT29-MTX cells). Many milk-derived peptides from whey and casein and wheat were taken into the investigation. These peptides bind the opioid receptors and therefore are more probable to have a role in mucus secretion. Human BCM-5 and alpha-lactorphin were chosen to investigate the MUC5AC (mucin 5AC gene) in the HT29-MTX cell line. The alpha-lactalbumin derived α-lactorphin demonstrated an increase in the expression of this gene from 4 h to 1 day. The increase in the expression of this gene was 1.6-fold over basal level expression, although differences were statistically different only after 24 h of exposure. Therefore, it is concluded that α-lactorphin has enough scope in up-regulating the expression of mucin gene and therefore in the maintenance of mucosal immunity (Martínez-Maqueda et al. 2012).

References

Belem M, Gibbs B, Lee B (1999) Proposing sequences for peptides derived from whey fermentation with potential bioactive sites. J Dairy Sci 82(3):486–493

Bläckberg L, Hernell O (1980) isolation of lactoferrin from human whey by a single chromatographic step. FEBS Lett 109(2):180–183

Bordenave S, Sannier F, Ricart G et al (1999) Continuous hydrolysis of goat whey in an ultrafiltration reactor: generation of alpha-lactorphin. Prep Biochem Biotechnol 29:189–202

Day A, Freer R, Liao C (1981) Morphiceptin beta-casomorphin (1-4) amide: a peptide opioid antagonist in the field stimulated rat vas deferens. Res Commun Chem Pathol Pharmacol 34 (3):543–546

Garg S, Nurgali K, Mishra KV (2016) Food proteins as source of opioid peptides-A rev. Curr Med Chem 23:893–910

Martínez-Maqueda D, Miralles B, De Pascual-Teresa S et al (2012) Food-derived peptides stimulate mucin secretion and gene expression in intestinal cells. J Agric Food Chem 60(35):8600–8605

Meisel H, FitzGerald RJ (2000) Opioid peptides encrypted in intact milk protein sequences. Br J Nutr 84(S1):27–31

Mullally MM, Meisel H, FitzGerald RJ (1997) Identification of a novel angiotensin-I-converting enzyme inhibitory peptide corresponding to a tryptic fragment of bovine b-lactoglobulin. FEBS Lett 402:99–101

Nurminen ML, Sipola M, Kaarto H et al (2000) α-Lactorphin lowers blood pressure via radiotelemetry in normotensive and spontaneously hypertensive rats. Life Sci 66:1535–1543

Pandey M, Kapila S, Kapila R et al (2018) Evaluation of the osteoprotective potential of whey derived-antioxidative (YVEEL) and angiotensin-converting enzyme inhibitory (YLLF) bioactive peptides in ovariectomised rats. Food Funct 9(9):4791–4801

Sipola M, Finckenberg P, Korpela R et al (2002) Effect of long-term intake of milk products on blood pressure in hypertensive rats. J Dairy Res 69:103–111

Tani F, Iio K, Chiba H et al (1990) Isolation and characterization of opioid antagonist peptides derived from human lactoferrin. Agric Biol Chem 54(7):1803–1810

Teschemacher H, Koch G, Brantl V (1997) Milk protein-derived opioid receptor ligands. Pept Sci 43(2):99–117

Casoxins

5

Abstract

Casoxins are food exorphins released from the κ-casein of milk protein. The genetic variants of κ-casein are AA, AB, BB, AC, BC, and AE. AA, AB, and BC κ-casein variants show casein micelle with diameters less than 200 nm and remaining variants form greater than 200 nm. Many reports establish BB fraction with increased and stronger gelling properties and better suitability for the preparation of cheeses. Casoxin A (YPSYGLN), B (YPYY), C (YIPIQYVLSR), and D (YVPFPPF) are protein fragments from 35–41, 57–60, 25–34, and 158–164, respectively, of κ-casein milk protein. Casoxins act as an opioid antagonist for the opioid receptors. Casoxin A has selectivity for all the three opioid receptors (μ/δ/κ). Casoxin D has selectivity for either μ or δ-opioid receptors. Casoxin B and C have an affinity for the only μ-opioid receptor. Casoxins have been reported to reverse the inhibitory action in the contractions induced by morphine in guinea pigs and mice in vitro. These exorphins have potential immunomodulatory roles in newborns that may be of the clinical significance in allergic diseases. It is suggested that these exorphins may have tremendous scope in the food and pharmacological industry.

5.1 Introduction

Cow milk is a source of proteins in the form of caseins and whey proteins. Caseins are further composed of alpha (α)-casein, beta (β)-casein, and kappa (κ)-caseins. Casein (κ) in cow milk has various genetic variants including AA, AB, BB, AC, BC, and AE. The variants of κ-casein (AA, AB, and BC) show casein micelle with diameters less than 200 nm, and the other genetic variants like AA, AC, and AE form these micelles greater than 200 nm (Bijl et al. 2014; Lodes et al. 1996). Many reports show the cow milk with more BB fraction of κ-casein having increased and stronger gelling properties and better suitability for the preparation of cheeses compared to

M. R. Ul Haq, *Opioid Food Peptides*, https://doi.org/10.1007/978-981-15-6102-3_5

milk with other variants (Walsh et al. 1996; Ikonen et al. 1999; Auldist et al. 2002; Amenu et al. 2006; Amenu and Deeth 2007; Jõudu et al. 2009). However, the cow milk comprised of BB type of κ-casein has a decreased coagulation time. The milk with the same κ-casein variant demonstrated enhances curd firmness in comparison to the milk with AA type of κ-casein (Ng-Kwai-Hang 1998). The cow milk with improved coagulation (rennet) attributes is because of the increased BB variant of κ-casein (Bijl et al. 2014; Walsh et al. 1996). The cow milk containing a lower ratio of casein/κ-casein supports the bridging of proteins with calcium that results in quicker coagulation and a stronger gel (Robitaille et al. 1993). Moreover, the cow milk containing more BB type κ-casein that is used for cheese preparation shows higher fat recovery and yield in comparison to that of the AA variant (Walsh et al. 1996). Furthermore, the cow milk with BB type of κ-casein is commercially significant from the cheese manufacturing point of view. This may be due to the size of micelles that are formed from this variant. All these caseins on gastrointestinal digestion yield amino acids that form the amino acid pool of the cells in humans. During gastrointestinal digestion, these proteins yield caseinomacropeptide that is accountable for enhanced effectiveness of digestion, prevention against neonate allergies to consumed proteins, and protection against gastric pathogens. However, some peptides are formed as such and are absorbed in the intestine and reach the circulation, in some cases the blood–brain barrier (BBB) and show effects in the brain. The opioid alkaloids like morphine and others have been demonstrated in controlling the pain sensations, however, these suffer from numerous side effects including dizziness, sedation, tolerance, addiction, nausea, constipation, vomiting, and respiratory depression. The opioid peptides derived from animal and plant proteins are safe alternatives as these do not show these side effects and have various options for their production.

5.2 Structure of Casoxins

The food-derived opioid peptides are also known as food exorphins. They are better regarded as the atypical opioid peptides in contrast to the typical opioid peptides of endogenous origin. For the first time, peptides were first extracted in 1979 from food sources and demonstrated opioid activity like morphine (Zioudrou et al. 1979). They mostly possess tyrosine at the N-terminal (Janecka et al. 2004). They further contain another aromatic amino acid at the third or fourth position. This arrangement or pattern of amino acids in these peptides forms a structural motif that fits in the binding site of opioid receptors (Teschemacher et al. 1997; Pihlanto-Leppala 2000). Although, some opioid peptides lack tyrosine residue at the amino terminus, nevertheless, they show opioid activity. For example, an opioid peptide (RYLGYLE) released from bovine α-casein lacks tyrosine at the amino-terminal, demonstrates opioid activity. Generally, reports have shown that animal-derived opioid has more selectivity towards the μ-opioid receptors (MOR). The plant food-derived opioid peptides demonstrate more specificity for the δ-opioid receptors (DOR) (Yoshikawa et al. 2003) excluding soymorphins. The animal food, i.e. milk is considered as a

Table 5.1 The amino acid sequence of casoxins

Casoxins	AA	Amino acid sequence
Casoxin A	07	Nt....YPSYGLN.......Ct
Casoxin B	04	Nt....YPYY...........Ct
Casoxin C	10	Nt....YIPIQYVLSR...Ct
Casoxin D	07	Nt....YVPFPPF........Ct

Nt amino-terminal, *Ct* carboxyl terminal, *AA* No. of amino acids

complete diet and contains caseins (α, β, and κ-casein) and whey proteins like lactalbumin and lactoglobulin. The enzymatic hydrolysis or fermentation releases opioid peptides from these proteins that have demonstrated opioid activity. The opioid peptides from β-casein and whey proteins have been discussed comprehensively in the previous chapters. In this chapter, the opioid peptides from the kappa-casein of bovine milk and α-casein of human milk will be discussed. These peptides are called casoxins and are A, B, C, and D type. The amino acid sequence of casoxins is given in Table 5.1 and the structure of these bioactive peptides is shown in Fig. 5.1.

5.3 Classification of Casoxins

As aforesaid these peptides are either released from the kappa-casein of bovine milk or alpha casein of human milk. These peptides are classified into four types named casoxin A, B, C, and D as depicted in Table 5.2.

Casoxin A
This peptide is derived from the milk of bovine kappa-casein. It is an opioid peptide having seven amino acids with a sequence of YPSYGLN from the fragment 35 to 41. It has a molecular formula ($C_{47}H_{61}N_9O_{14}$) and molecular weight 976 g/mol. It acts as an opioid antagonist and has opioid receptor selectivity for all the three opioid receptors like MOP, KOP, and DOP. However, the selectivity is far greater towards MOP compared to DOP or KOP.

Casoxin B
This peptide is also produced from the milk of bovine kappa-casein. It is an opioid peptide having four amino acids with a sequence of YPYY from the fragment 57 to 60. It has a molecular formula ($C_{32}H_{36}N_4O_8$) and molecular weight 604.6 g/mol. It acts as an opioid antagonist and has opioid receptor selectivity for only MOP.

Casoxin C
This peptide is also produced from the milk of bovine kappa-casein. It is an opioid peptide having four amino acids having a sequence of YIPIQYVLSR from the fragment 25 to 34. It has a molecular formula ($C_{60}H_{94}N_{14}O_{15}$) and molecular weight 1251.5 g/mol. It acts as an opioid antagonist and has opioid receptor selectivity for only MOP (Chiba et al. 1989; Takahashi et al. 1997).

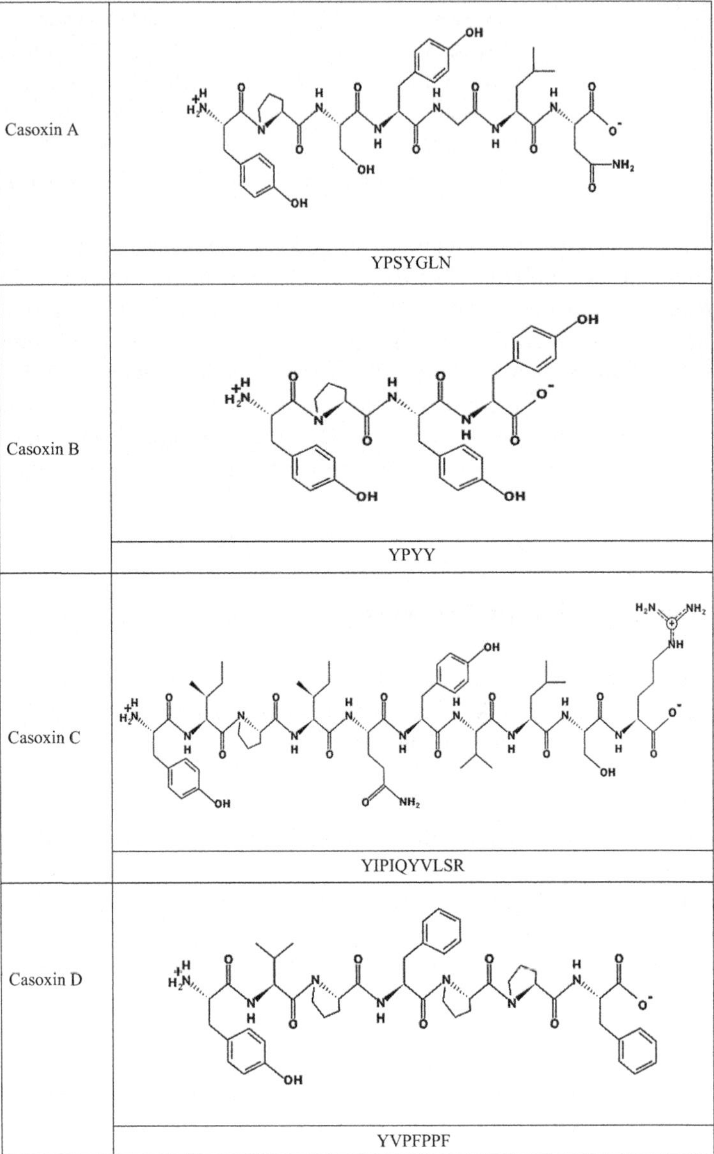

Fig. 5.1 Structure of casoxins

Casoxin D

This peptide is also produced from the milk of bovine alpha (S1)-casein. It is an opioid peptide having four amino acids having a sequence of YVPFPPF from the fragment 158 to 164. It has a molecular formula ($C_{47}H_{59}N_7O_9$) and molecular weight

Table 5.2 Classification of casoxins

Casoxins	Sequence	PP	OR	F	A/AN	ORS	Mr (g/mol)	MF
Casoxin A	YPSYGLN	β-casein	Bovine	35–41	AN	$\mu \gg \delta,\kappa$	976	$C_{47}H_{61}N_9O_{14}$
Casoxin B	YPYY	β-casein	Bovine	57–60	AN	μ	604.6	$C_{32}H_{36}N_4O_8$
Casoxin C	YIPIQYVLSR	β-casein	Bovine	25–34	AN	μ	1251.5	$C_{60}H_{94}N_{14}O_{15}$
Casoxin D	YVPFPPF	α-casein	Human	158–164	AN	μ/δ	866	$C_{47}H_{59}N_7O_9$

PP parent protein, *OR* organism, *F* fragment, *A* agonist, *AN* antagonist, *Mr* molecular weight, *MF* molecular formula

866 g/mol. It acts as an opioid antagonist and has opioid receptor selectivity for only MOP or DOP (Yoshikawa et al. 1994).

5.4   Opioid Activity of Casoxins

When the bovine κ-casein was digested with pepsin and trypsin, two peptides were purified that were established to show opioid antagonistic properties (Yoshikawa et al. 1986; Chiba et al. 1989). Depending on opioid properties, numerous peptides were synthesized through chemical synthesis and many of these peptides showed opioid activity. One of these opioid peptides was represented by casoxin C with 10 amino acid residues. Besides this, the other opioid peptide is composed of ten amino acid residues with modified C-terminal Tyr residue. Seemingly, the latter gets methoxylated through any of the purification steps. The opioid peptide casoxin (1–6) demonstrated the antagonistic activity for μ-receptors on GPI preparation and the activity of a κ-agonist on the preparation made from rabbit vas deferens, but not the effect of a δ agonist on the preparation made from mouse vas deferens preparation. When human α or κ-casein was digested with pepsin and chymotrypsin (Johnsen et al. 1994), a peptide was purified from α-casein with an opioid activity that depicted opioid antagonistic activity (Yoshikawa et al. 1994). This peptide was derived from human αs1 casein from 158–164 and named casoxin D (1–7). This peptide demonstrated a very low affinity towards the receptor and antagonized μ-agonist in the preparation made from guinea-pig ileum in addition to the inhibitory activity of a δ agonist in the preparation made from mouse vas deferens. It was also reported that all showed antagonistic selectivity almost two orders of magnitude lower compared to the naloxone. The casoxin (1–6) is considered as the μ and κ-receptor antagonist with low affinity. The selectivities and opioid potencies for synthetic analogs and casoxin D seem to be in range. Nevertheless, the fragments from k-casein demonstrated negligible activities (Yoshikawa et al. 1986). Earlier Yoshikawa and coworkers in 1986 isolated a bioactive peptide with antagonist opioid activity. The amino acid sequence of the peptide was Ser-Arg-Tyr-Pro-Ser-TyrOCH$_3$ and the peptide was regarded as casoxin 4. The bioactive peptide was a fragment from 33 to 38 amino acid residue of κ-casein with carboxyl ester at C-terminal. The isolated peptide showed activity (IC$_{50}$) value of 1 nM in presence of naloxone (15 mM) when observed through radioreceptor assay. The researchers could not find any agonistic activity through any of the analyzed assays. On the tissue preparations made from rabbit vas deferens and guinea-pig ileum, the peptide showed antagonistic activities in presence with dynorphin A and morphiceptin, respectively. When the tissue preparations were made from mouse vas deferens, the bioactive peptide could not demonstrate the antagonistic activity with [Leu] enkephalin. These studies established that this peptide was antagonist selective for the μ and κ-opioid receptors (Yoshikawa et al. 1986). Casoxin C, a physiologically active peptide was isolated from the kappa-casein hydrolysates that were digested with trypsin. It was found to be an anti-opioid bioactive peptide when its activity was subjected to guinea-pig ileum (GPI). Results have further established that this peptide demonstrated contractions of the ileal strips. The latter phenomenon was observed as biphasic with rapid and slow modules. The contractile contour was quite similar to C3a (70–77) (human complement), which is a carboxyl end of an octapeptide of this complement protein. Although the activities showed less potency quantitatively, nevertheless the qualitative indications demonstrated similar biological activities like that of C3a. The in silico methods have established a sequence homology between this milk-derived bioactive peptide and complement protein (C3a). Moreover, the rapid contraction was facilitated through histamine production, while the alternative phenomenon (slow contraction) was intervened through prostaglandin E2-synthesis. These observations were taken through the analysis of the application of numerous pharmacological antagonists and inhibitors. This kappa-casein derived peptide has been found to have an affinity for C3a receptors with IC_{50} values of 40 mM observed through the radioreceptor method. Additionally, this peptide demonstrated phagocyte-stimulating activities. In conclusion, this was the first milk-derived bioactive peptide found to act through complement C3a receptors. Therefore these observations establish that this milk-derived peptide may have a potential role in the development of immune functions in humans (Takahashi et al. 1997).

5.5 Production of Casoxins

Casoxin C was isolated from the bovine milk kappa-casein using well-established hydrolysis protocols through the bioanalytical approach. The researchers weighed 3 g of bovine κ-casein and dissolved in water (300 ml). The pH was adjusted to 1.4 through the incremental addition of concentrated HCl. The hydrolysis of the bovine κ-casein was performed at 37 °C for 5 h. The highly purified grade of enzyme porcine pepsin weighing 60 mg in the ratio of 1:60,000 was taken for this purpose. The hydrolysate made in the preceding protocol was lyophilized and then extracted with the use of solvents like chloroform: methanol (65: 35). The latter extract was further dried through evaporation and the resultant was reconstituted in distilled at pH 7 with 10N KOH. Differential centrifugation (10,000 rpm for 15 min) was exploited to remove the insoluble material. The supernatant was loaded on the octadecyl silica (ODS) column. Gradient elution was performed from zero to 50% for 50 min of acetonitrile (CAN) having 0.1% trifluoroacetic acid (TFA). The individual fractions were taken and then dried with the help of a centrifugal concentrator and active fractions were purified on the cyanopropyl silica (CN) column. While using this protocol of κ-casein, pepsin hydrolysis, and then analysis on reverse-phase HPLC, the researchers were able to isolate a bioactive peptide with antagonist opioid activity. The amino acid sequence of the peptide was Ser-Arg-Tyr-Pro-Ser-TyrOCH$_3$ and the peptide was regarded as casoxin 4. The bioactive peptide was a fragment from 33 to 38 amino acid residue of κ-casein with carboxyl ester at C-terminal. The analysis of the amino acid of the isolated bioactive peptide was done on a Hitachi 835 amino acid analyzer. The hydrolysis was performed at 110 °C for 24 h in 6N HCl. The sequence of the amino acid of the isolated bioactive peptide was performed on 470A gas-phase protein sequencer (Yoshikawa et al. 1986). The other investigators used the same hydrolysis protocols and sequencing methods for the isolation of casoxins from κ-casein of bovine milk. The results indicated a bioactive peptide fragment corresponding to 25–34 of bovine κ-casein with a sequence Tyr-Ile-Pro-Ile-Gln-Tyr-Val-Leu-Ser-Arg. The peptide was regarded as casoxin C and showed activity at 5 μM through GPI assay (Chiba et al. 1989). Moreover, recombinant DNA technology was exploited by the production of casoxin. In this perspective, Kato and coworkers in 1995 tried to produce the casoxin D (Bradykinin agonist), the peptide is released from the human alpha-casein, by *Bacillus brevis*. The investigators were successful in producing this opioid peptide while exploiting *Bacillus brevis* as a host. The DNA sequence that codes for this opioid peptide was tandemly ligated. Since the expression of peptide had to be done, therefore, an expression vector was selected for the study. In addition to the target DNA sequence, a fusion protein sequence for an epidermal growth factor (EGF) was also chosen. Therefore an expected fusion protein casoxin-D-EGF was prepared and *B. brevis* was transformed with this expression vector. The results indicated the production of this protein (0.5 g/l) in the culture supernatant. This protein was then purified with a salting-out method using ammonium sulfate. The conjugate (peptide-EGF) was hydrolyzed with the help of two types of proteases. Reverse phase-HPLC was used for the identification of this casoxin D (Kato et al. 1995).

5.6 Biological Activities of Casoxins

The research on the opioid peptides derived from the κ-casein fraction of cow milk has shown their role in contractility of the intestine and the transit time of gut in mice models. The use of conventional opioids as the analgesic agents has the side effects of induction of constipation in human subjects. In this perspective, rat, mouse, and guinea pig were taken for the study, and their ileal sections were dissected and stimulated electrically. This was followed by the induction of inhibition in vitro with the morphine. The latter phenomenon (morphine inhibition) inducted in guinea pig and mouse was in turn blocked by opioid subtype antagonists. Poly R-478 (polymeric dye), casoxin 4, and lactoferroxin A (opioid antagonists) were given orally and checked for blocking action induced by morphine in animal models by either single or double gavage methods. The results indicated that the tissue taken from the guinea pig was more sensitive to the inhibitory action of morphine in comparison to the rat or mouse tissue. The mouse showed remarkably δ and κ-opioid receptor actions having a lesser μ-opioid receptor element. Moreover, the observations demonstrated that both guinea pig and mouse tissues were sensitive to casoxin-4 (derived from κ-casein) antagonism, the inhibitory action on the contraction of ileum previously induced by morphine. Furthermore, contrariwise to naloxone, comparatively high doses (oral) of lactoferroxin and casoxin 4 (μ-opioid receptor antagonists), that were applied before and after morphine introduction was found to be unable to antagonize the inhibitory action of morphine for gut transit. In conclusion, it may be put forward that milk protein-derived peptide (casoxin 4) may reverse inhibitory action in the contraction induced by morphine in guinea pigs and mice in vitro. The same opioid peptide does not affect the morphine inhibitory activity in mouse small intestinal transit through oral administration (Patten et al. 2011). As aforementioned in the previous chapters the milk is a complete diet for infants and a part of the food for the adults also. This milk serves the nutritional aspects, however, during a few decades; a physiological perspective of this milk has been highlighted. The latter phenomenon is related to the production of the bioactive peptides in the gut that regulate various physiological processes. Keeping the above regards under consideration various researchers have explored numerous aspects in this field. In a similar type of research, Fiedorowicz and coworkers tried to assess the effect of milk-derived bioactive peptides, bovine and human BCM-7 (agonists) and kappa-casein derived casoxin D and 6 on the proliferation of peripheral blood mononuclear cells (PBMCs) and secretion of immunological molecules, i.e. cytokines. The proliferation of these cells was assessed through the BrdU test that checks the DNA synthesis activity. Moreover, the water soluble tetrazolium salts (WST-1) test that checks the activity of mitochondrial dehydrogenase enzymes was also performed. The latter test checks the metabolic activity of the mitochondria viability of cells. Cell proliferation in addition to indirect cell death may be assessed by this test on a large scale with the help of microtiter plates. Enzyme-linked immunosorbent assay (ELISA) was exploited for the assessment of the levels of IL-4, IL-8, IL-13, and IFN-Y on incubation of these cells with milk-derived opioid peptides. WST-1 test showed that there was no significant change in the enzymatic activity in these cells with these peptides. Nevertheless, the BrdU test indicated the changes in the proliferation of PBMCs. Furthermore, alterations in the levels of IL-4 and IL-13 by PBMCs were observed on the application of agonists in addition to the changes in the levels of IFN-Y. These observations, therefore, provide valuable information on the immunomodulatory activity of food exorphins that could be of the clinical significance more specifically related to the allergic diseases manifesting in newborns (Fiedorowicz et al. 2011). Moreover, a patent (PCT/EP2008/007842) was filed in September 2008 and published in April 2009 (WO 2009/046829 A2) claims that the peptide casoxin D with a sequence (YVPFPPF) may prove as a therapeutic agent against the prophylaxis. It may, therefore, have a potential for the treatment of cancer fibrotic diseases, inflammatory responses, and neurodegenerative complications. Moreover, it may be beneficial in ameliorating the conditions like autoimmune and infectious diseases in addition to the lung and cardiovascular ailments. Therefore, the patent claims that this peptide may have tremendous scope in the pharmacological industry for the preparation of formulations, especially in liquid buffer or lyophilisate. The formulations may also be given through mother's milk supplemented with this peptide at doses greater than already present lyoprotectant, cryoprotectant, and excipient (Dorian 2009).

References

Amenu B, Deeth H (2007) The impact of milk composition on Cheddar cheese manufacture. Aust J Dairy Technol 62:171–184

Amenu B, Cowan T, Deeth H et al (2006) Impacts of feeding system and season on milk composition and Cheddar cheese yield in a subtropical environment. Anim Prod Sci 46:299–306

Auldist M, Mullins C, O'Brien B et al (2002) Effect of cow breed on milk coagulation properties. Milchwissenschaft 57:140–143

Bijl E, De Vries R, Van Valenberg H et al (2014) Factors influencing casein micelle size in milk of individual cows: genetic variants and glycosylation of k-casein. Int Dairy J 34:135–141

Chiba H, Tani F, Yoshikawa M (1989) Opioid antagonist peptides derived from κ-casein. J Dairy Res 56:363–366

Dorian B (2009) Use of a peptide as a therapeutic agent. PCT/EP2008/007842

Fiedorowicz E, Jarmołowska B, Iwan M et al (2011) The influence of μ-opioid receptor agonist and antagonist peptides on peripheral blood mononuclear cells (PBMCs). Peptides 32(4):707–712

Ikonen T, Ahlfors K, Kempe R et al (1999) Genetic parameters for the milk coagulation properties and prevalence of noncoagulating milk in Finnish dairy cows. J Dairy Sci 82:205–214

Janecka A, Fichna J, Janecki T (2004) Opioid receptors and their ligands. Curr Top Med Chem 4:1–17

Johnsen LB, Rasmussen LK, Petersen TE et al (1994) Characterization of three types of human alpha s1-casein mRNA transcripts. Biochem J 309:237–242

Jõudu I, Henno M, Värv S et al (2009) The effect of milk proteins on milk coagulation properties in Estonian dairy breeds. Vet Zootec 46:14–19

Kato M, Fujiwara Y, Okamoto A et al (1995) Efficient production of casoxin D, a bradykinin agonist peptide derived from human casein, by bacillus brevis. Biosci Biotechnol Biochem 59 (11):2056–2059

Lodes A, Krause I, Buchberger J et al (1996) The influence of genetic variants of milk proteins on the compositional and technological properties of milk. Casein micelle size and the content of non-glycosylated kappa-casein. Milchwissenschaft 51:368–373

Ng-Kwai-Hang K (1998) Genetic polymorphism of milk proteins: relationships with production traits, milk composition and technological properties. Can J Anim Sci 78:131–147

Patten GS, Head RJ, Abeywardena MY (2011) Effects of casoxin 4 on morphine inhibition of small animal intestinal contractility and gut transit in the mouse. Clin Exp Gastroenterol 4:23–31

Pihlanto-Leppala A (2000) Bioactive peptides derived from bovine whey proteins: opioid and ACE-inhibitory peptides. Trends Food Sci Technol 11(9):347–356

Robitaille G, Ng Kwai Hang KF, Monardes IIG (1993) Effect of k-casein glycosylation on cheese yielding capacity and coagulating properties of milk. Food Res Int 26:365–369

Takahashi M, Moriguchi S, Suganuma H et al (1997) Identification of casoxin c, an ileum-contracting peptide derived from bovine kappa-casein, as an agonist for c3a receptors. Peptides 18(3):329–336

Teschemacher H, Koch G, Brantl V (1997) Milk protein-derived opioid receptor ligands. Pept Sci 43(2):99–117

Walsh EM, McSweeney PLH, Fox PF (1996) Use of antibiotics to inhibit non-starter lactic acid bacteria in Cheddar cheese. Int Dairy J 6:425–431

Yoshikawa M, Tani F, Ashikaga T et al (1986) Purification and characterization of an opioid antagonist from a peptide digest of bovine κ-casein. Agric Biol Chem 50:2951–2954

Yoshikawa M, Tani F, Shiota H et al (1994) Casoxin D, an opioid antagonist ileum-contracting/vasorelaxing peptide derived from human αs1-casein. In: Brantl V, Teschemacher H (eds) β-casomorphins and related peptides: recent developments. Wiley, Weinheim, pp 43–48

Yoshikawa M, Takahashi M, Yang S (2003) Delta opioid peptides derived from plant proteins. Curr Pharm Des 9(16):1325–1330

Zioudrou C, Streaty RA, Klee WA (1979) Opioid peptides derived from food proteins. J Biol Chem 254(7):2446–2449

Gluten Exorphins

6

Abstract

Gluten is a complex structural protein in wheat, composed of glutenins and gliadins. The former is soluble in dilute acids, whereas the latter is alcohol-soluble protein. Gluten exorphins A4, A5, B4, B5, and C5 are released from the high molecular weight glutenins and gliadorphin-7 is released from the α-gliadin of wheat. A5 is a pentapeptide (GYYPT) opioid agonist and exhibits selectivity for δ- and μ-opioid receptors. A4 is a tetrapeptide (YGGP) opioid agonist that shows selectivity for δ- and μ-opioid receptors. B5 is a pentapeptide (YGGWL) opioid agonist that possesses an affinity for the δ-opioid receptor. B4 is a tetrapeptide opioid (YGGW) agonist that shows an affinity for the δ-opioid receptor. C5 is a pentapeptide (YPISL) opioid agonist that is selective for the μ-opioid receptor. Gliadorphin-7 is a heptapeptide (YPQPQPF) opioid agonist released from α-gliadin of wheat. Many reports have established the increased production and absorption of wheat-derived exorphins with increased risk of neurological manifestations. Moreover, there are studies available that correlate the consumption of these exorphins with the celiac disease. A possible role of these exorphins is suggested in pituitary function through conventional opioid receptors. These peptides have a role in prolactin secretion, modification of brain neurotransmitter secretion, hormonal and gastrointestinal functions. Gluten exorphin A5 is reported to modulate pain senses, observable behavior, learning, and memory processes in animal models.

6.1   Introduction

Gluten is a complex protein composed of glutenins and gliadins. It is a structural protein in wheat. The glutenin fraction is soluble in dilute acids and the polymers of individual proteins. The gliadins, on the other hand, are alcohol soluble derived from wheat (Tatham and Shewry 2008). For the first time, Zioudrou and coworkers in

1979 isolated exorphins from wheat after pepsin hydrolysis because they are derived exogenously from food sources and demonstrate morphine-like activity (Zioudrou et al. 1979). Consequently, other workers isolated five gluten exorphins from wheat including gluten exorphin A4 (Gly-Tyr-Tyr-Pro), A5 (Gly-Tyr-Tyr-Pro-Thr), B4 (Tyr-Gly-Gly-Trp), B5 (Tyr- Gly-Gly-Trp-Leu), and C (Tyr-Pro-Ile-Ser-Leu). The opioid activity of these peptides was confirmed on various pharmacological techniques including electrically stimulated mouse vas deferens (MVD), guinea pig ileum (GPI), radio-receptor assay. These assays assess the selectivity of these opioid peptides to their opioid receptors. The gluten exorphin (A/B) usually shows more selectivity towards the δ-opioid receptor. The gluten exorphin demonstrates more selectivity towards the μ-opioid receptor. The gluten exorphin B5 among all the gluten exorphins showed the highest potency for the opioid receptors. The gluten exorphins A/B are derived from the gluten and the gliadorphins 5 and 7 are generated from gliadorphin of wheat (Fukudome and Yoshikawa 1992, 1993).

6.2   Structure of Gluten Exorphins

The exorphins from the wheat are either derived from high molecular weight glutenin or alpha-gliadin. The gluten exorphins (A/B/C) are derived from glutenin and the gliadorphin is derived from the alpha-gliadin. The A-type gluten exorphins both possess common four amino acids (GYYP) from the amino-terminal. The B type gluten exorphins both demonstrate the common first four amino acids (YGGW) from amino-terminal. The gluten exorphin C and gliadorphin both show tyrosine at the N-terminal that is essential for the demonstration of opioid activity. These wheat-derived opioid peptides are derived from the proteins during gastrointestinal digestion in vivo, simulated gastrointestinal digestion (SGID) in vitro, or with bacterial fermentation. Reports have established that the patients with neurological manifestations like autism and schizophrenia have often leaky gut that allows these peptides to be absorbed more readily, cross the blood–brain barrier (BBB), and then bind opioid receptors in the brain to show various manifestations (Cade et al. 2000). There are various pieces of evidence in favor of this hypothesis and others are against this hypothesis. The formulation of wheat and gluten-free diet is based on this hypothesis (Christison and Ivany 2006). The structures of gluten exorphins are shown in Fig. 6.1.

6.3   Classification of Gluten Exorphins

Based on the number, amino acid sequence, and source of proteins, these wheat-derived exorphins have six types as shown in Table 6.1

Gluten Exorphin A4
This is a tetrapeptide derived from the wheat. The high molecular weight protein glutenin is the source of this opioid peptide. It has a sequence GYYP and has far

Fig. 6.1 Structures of gluten exorphins

Gluten Exorphin A4	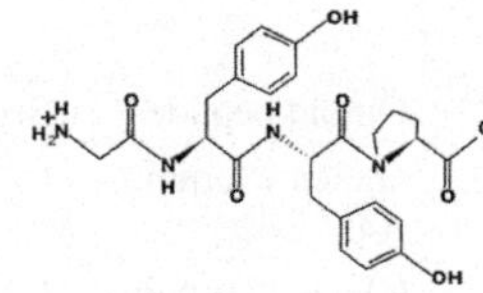 GYYP
Gluten Exorphin A5	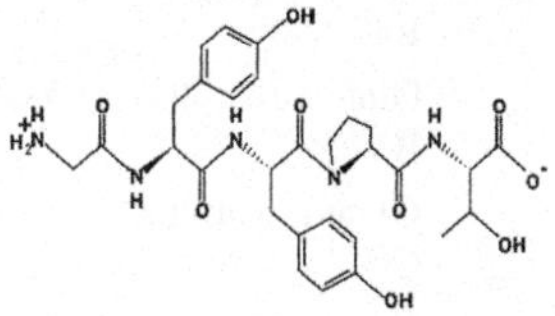 GYYPT
Gluten Exorphin B4	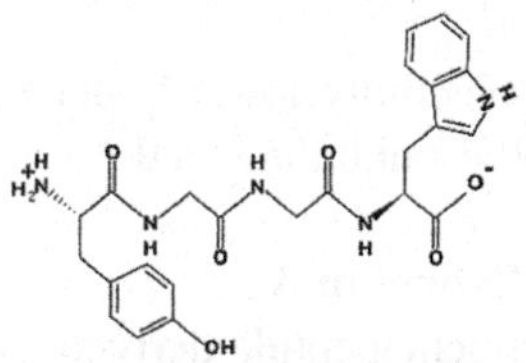 YGGW
Gluten Exorphin B5	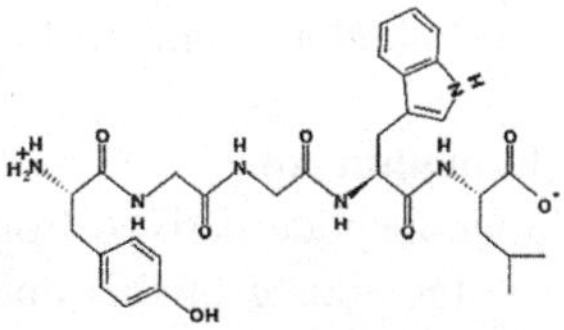 YGGWL
Gluten Exorphin C	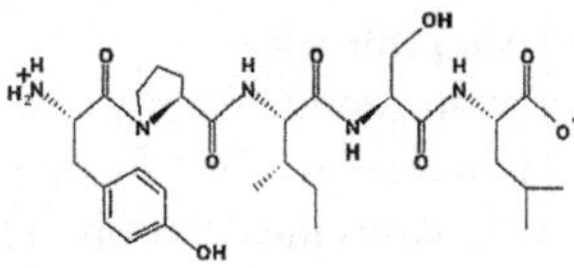 YPISL
Gliadorphin-7	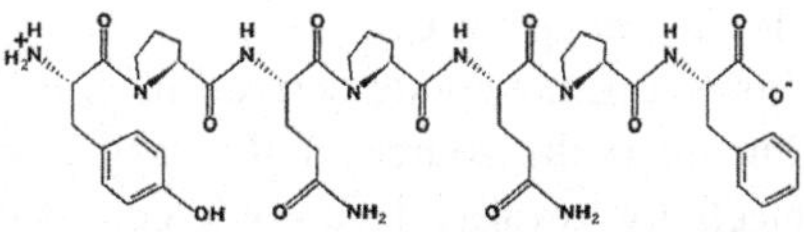 YPQPQPF

Table 6.1 Classification of gluten exorphins

Protein source	Opioid peptide	Sequence	ORS	A/ AN	Mr (g/mol)	MF
Glutenin	Gluten exorphin A4	GYYP	$\mu \ll \delta$	A	498.2	$C_{25}H_{28}N_4O_6$
	Gluten exorphin A5	GYYPT	$\mu \lll \delta$	A	599.6	$C_{29}H_{37}N_5O_9$
	Gluten exorphin B4	YGGW	δ	A	481.1	$C_{24}H_{27}N_5O_9$
	Gluten exorphin B5	YGGWL	δ	A	594.7	$C_{30}H_{38}N_6O_7$
	Gluten exorphin C	YPISL	$\mu < \delta$	A	591.7	$C_{29}H_{45}N_5O_8$
α-gliadin	Gliadorphin-7	YPQPQPF	δ	A	875.4	$C_{43}H_{57}N_9O_{11}$

ORS opioid receptor selectivity, *A* agonist, *AN* antagonist, *Mr* molecular weight, *MF* molecular formulae

greater selectivity towards delta compared to the mu-opioid receptor (Yoshikawa et al. 2003; Fukudome and Yoshikawa 1992).

Gluten Exorphin A5

This is a pentapeptide derived from the wheat. The high molecular weight protein glutenin is the source of this opioid peptide. It has a sequence GYYPT and has far-far greater selectivity towards delta compared to the mu-opioid receptor (Yoshikawa et al. 2003; Fukudome and Yoshikawa 1992).

Gluten Exorphin A4

This is a tetrapeptide derived from the wheat. The high molecular weight protein glutenin is the source of this opioid peptide. It has a sequence YGGW and has selectivity towards the delta-opioid receptor. It acts as an opioid agonist (Yoshikawa et al. 2003; Fukudome and Yoshikawa 1992).

Gluten Exorphin B5

This is a pentapeptide derived from the wheat. The high molecular weight protein glutenin is the source of this opioid peptide. It has a sequence YGGWL and has selectivity towards the delta-opioid receptor. It acts as an opioid agonist (Yoshikawa et al. 2003; Fukudome and Yoshikawa 1992).

Gluten Exorphin C5

This is a pentapeptide derived from the wheat. The high molecular weight protein glutenin is the source of this opioid peptide. It has a sequence YPISL and has selectivity towards delta compared to the mu-opioid receptor. It acts as an opioid agonist (Yoshikawa et al. 2003; Fukudome and Yoshikawa 1993).

Gliadorphin-7

This is a heptapeptide derived from the wheat. The wheat alpha-gliadin is the source of this opioid peptide. It has a sequence YPQPQPF and has selectivity towards the delta-opioid receptor. It acts as an opioid agonist.

6.4   Opioid Activity of Gluten Exorphins

Endorphins are the molecules that are synthesized inside the human body and act as ligands for the opioid receptors. These receptors are located in the central and peripheral nervous system, gastrointestinal tract, and some immune cells. These endorphin peptides have a common amino-terminal with a sequence Tyr-Gly-Gly-Phe-(Met or Leu), which is denoted as the opioid motif. This sequence in the opioid peptide is followed by extensions of 5–31 amino acid residues at the carboxyl-terminal (Aldrich and McLaughlin 2009). There is a tremendous scope of μ-opioid receptor (MOR) ligands or agonists including morphine. This is because these ligands have been exploited in the management of pain or analgesics. Therefore much development has been done in this area to produce analgesic effects through the utilization of μ-opioid receptor agonists via the central nervous system. In this perspective, there have been many trials that have well-established the generation of similar exogenous opioid peptides (Aldrich and McLaughlin 2009; Yamamoto et al. 2003) from wheat or casein through simulated gastrointestinal digestion or fermentation. These food-derived exorphins have similar structures mostly at the amino-terminal like the presence of tyrosine at the amino-terminal or bioactive clefts (Kitts and Weiler 2003; Yamamoto et al. 2003; Aldrich and McLaughlin 2009). The endogenous opioid peptides bind with various opioid receptors and regulate various physiological processes. The exorphins, on the other hand, are produced from various food sources like milk, wheat, rice, soybean, and spinach. These opioid peptides also bind these endogenous opioid receptors and thereby regulate various physiological processes. Under normal conditions, these opioid peptides may contribute towards the wellness of an individual. However, in abnormal conditions like the manifestation of leaky gut, these may be absorbed and transported in large quantities and thereby produce adverse effects (Liu and Udenigwe 2019). The opioid activity of food-derived peptides may be checked through various pharmacological techniques. The commonly employed assays include the naloxone reversible inhibition of enzyme activity (adenylate cyclase) and electrically stimulated contraction of tissue preparations of guinea pig ileum or mouse vas deferens (Sharma et al. 1975; Hughes et al. 1975). In addition to these assays, the radioreceptor assay (receptor binding assay) is also exploited for this purpose (Pert et al. 1973; Pert and Snyder 1973). The opioid receptors are categorized in the G protein-coupled receptors family (GPCRs). Whenever there is an activation of these receptors it causes activation and inactivation of K^+ and Ca^{2+} channels, respectively. Moreover, their activation also leads to the inhibition of adenylate cyclase that forms cyclic AMP from ATP (Satoh and Minami 1995). Many investigators have exploited the inhibition assay of adenylate cyclase through GPCRs for the determination of the opioid activity of exorphins released from casein and gluten (Zioudrou et al. 1979; Loukas et al. 1983; Yoshikawa et al. 2003). Whenever an exorphin binds to its corresponding opioid receptor in either brain membranes or hybrid cells, it results in the inhibition of adenylate cyclase (responsible for the formation of cAMP from ATP), this, in turn, decreases the concentration of cAMP that may be assessed through various assays (Sharma et al. 1975; Satoh and Minami 1995). Moreover, for the determination of the opioid activity of these food-derived exorphins, the tissues like guinea pig ileum or mouse vas deferens are first stimulated electrically, then opioid peptide binds to their corresponding receptors on this tissue and leads to the inhibition of the contractions (Janecka et al. 2004; Teschemacher 2003). The opioid peptides from the food sources bind selectively with MOP and DOP on the guinea pig ileum and mouse vas deferens preparations, respectively. In this way, these assays are employed from the assessment of MOP and DOP ligands from various exorphins (Janecka et al. 2004; Zioudrou et al. 1979). The receptor binding assay includes competition and saturation kinetics, and in the latter, the affinity of various food-derived exorphins for their corresponding receptors is evaluated. The former assay is used independently or subsequently to validate or confirm the results (Janecka et al. 2004). The disadvantage with the binding assay is the use of radioisotopes that are hazardous to use, need permissions, expensive, and further, their disposal is an issue. Moreover, the food-derived peptides that will alter the binding of a selected molecule allosterically may not essentially enhance the signaling potency. Furthermore, the generated data from various laboratories are difficult to compare due to changes in synaptosomal preparations, the doses, radioligand employed, and the assay used for reporting this binding. In conclusion, the bioassays that include tissue preparations and adenylate cyclase are more advantageous compared to receptor binding methods, this is due to the reason that these are more feasible, less expensive, and demonstrate the very low effect on the environment. Fukudome and Yoshikawa in 1992 isolated four opioid exorphins from the wheat gluten. The peptides were named as gluten exorphin 5 (Gly-Tyr-Tyr-Pro-Thr), gluten exorphin A4 (Gly-Tyr-Tyr-Pro), gluten exorphin B5 (Tyr-Gly-Gly-Trp-Leu), and gluten exorphin B4 (Tyr-Gly-Gly-Trp). The opioid activity of four isolated wheat-derived exorphins (A/B) was checked through GPI and MVD assays. In MVD assay, the mice were sacrificed and the vas deferens collected, the tension was applied to the isolated tissue (vas deferens) through suspension of 0.02 g weight in a Magnus tube that was supplemented with Mg^{2+} free Krebs-Ringer solution and stimulated. In the case of GPI assay, the ileum was taken and tension applied with 0.5 g weight in Krebs-Ringer solution and stimulated. The contraction was documented with a transducer (isometric). These experiments of assessment of opioid activity were performed in the presence of L-leucyl-L-leucine (2 mM) together with bestatin (30 μM), thiorphan (0.3 μM), and captopril (10 μM). The IC_{50} values are the concentration of these opioid peptides in the hydrolysates that inhibits the stimulated (electrically) muscle contraction by 50%. The assessment of opioid activities of these peptides through the MVD assay was done in the presence and absence of 10^{-6} μM naloxone. The opioid activities that were reversed on the addition of naloxone and blocked by pretreatment with this compound were regarded as opioid activities. The radioreceptor assays were done as per Pert and Snyder (1973) in presence of [^{3}H] DADLE (1 nM, 30.0 Ci/mmol) and [^{3}H] DAGO (1 nM, 47.8 Ci/mmol) for DOP and MOP and rat brain membrane. This radioisotope assay was performed for κ-receptors in presence of [^{3}H] EKC (1 nM, 24.5 Ci/mmol) and guinea pig cerebellum membrane. These opioid activity determination assays were done in the presence of inhibitors. The IC_{50} value is the concentration of the opioid peptide that will inhibit the binding of the labeled ligand by 50% (Fukudome and Yoshikawa 1992). Moreover, the researchers showed that the sequence of gluten exorphin A5 is located at 15 sites in the primary of glutenin that is high molecular weight protein found in wheat. This exorphin showed high selectivity towards delta opioid receptor. Moreover, the structure-function relationship established that this exorphin is unique in that it possesses glycine at the amino-terminal unconventional to the presence of tyrosine in other food-derived opioid peptides. The wheat gluten derived B4 exorphin was found to be the most potent among all purified exorphins in the experiment with IC_{50} values of 0.05 μM and 0.017 μM on GPI and MVD tissues, respectively (Fukudome and Yoshikawa 1992).

The same researcher continued the research in 1993 on isolation purification and opioid activity determination of wheat-derived opioid peptides. The investigators isolated novel wheat gluten derived exorphin with a sequence Tyr-Pro-Be-Ser-Leu after enzymatic hydrolysis using pepsin, trypsin, and chymotrypsin. The IC_{50} values determined were 40 μM and 13.5 μM when observed on the GPI and MVD tissues, respectively. The purified food-derived peptide was denoted as gluten exorphin C. This food-derived opioid peptide is unique in that it possesses a single tyrosine residue and that too at the amino-terminal. The research was further extended and the results indicated when the analogs of this exorphin Tyr-Pro-X-Ser-Leu were commercially or chemically synthesized to establish structure-function (activity) relationship. The synthesized peptides in which X amino acid was substituted by an aromatic or aliphatic hydrophobic amino acid demonstrated opioid activity. The chemically synthesized structural analogs in which X was substituted by hydrophobic (aliphatic) amino acid residues like leucine and valine were found to have lesser potency compared to gluten exorphin C. When X was substituted by hydrophobic (straight-chain) amino acid residues like N.leucine and N.valine were found to demonstrate same activity like that of gluten exorphin C. Similarly, when X was substituted by basic or hydrophilic amino acids like lysine, alanine, and threonine, the exorphin could not demonstrate any activity. The milk-derived BCM-7 and hemoglobin derived hemorphin-4 like analogs where X was substituted with aromatic amino acids phenylalanine or tryptophan demonstrated activity quite comparable to that of gluten exorphin C. Therefore it may conclude that Tyr-Pro-X-Ser-Leu analogs where X is substituted by a hydrophobic (aliphatic) or an aromatic amino acid showed opioid activity and the presence of hydrophobic amino acid residue at 3rd position remarkably altered their opioid activity. It was, therefore, suggested that the hydrophobicity of isoleucine at position 3 was significant for demonstration of the opioid activity of gluten exorphin C. This structure-functional association revealed that the second aromatic amino acid at 3rd position is not significant for the manifestation of the opioid activity of peptides with Tyr-Pro

Table 6.2 Opioid activity of gluten exorphin (Fukudome and Yoshikawa 1992, 1993)

Wheat exorphins	Opioid activity (IC_{50} µM)			Receptor affinities (IC_{50} µM)		
	Guinea pig ileum	Mouse vas deferens	µ/δ ratio	(^{3}H) DAGO	(^{3}H) DADLE	µ/δ ratio
Gluten exorphin A5	1000	60	16.7	700	1.5	4.67
Gluten exorphin A4	>1000	70	–	>1000	3.8	–
Gluten exorphin B5	0.05	0.017	2.9	0.045	0.005	9
Gluten exorphin B4	1.5	3.4	0.44	0.17	0.18	0.94
Gluten exorphin C	40	13.5	3.0	110	30	3.7

sequence (Fukudome and Yoshikawa 1992). The opioid activity of gluten exorphins is given in Table 6.2.

6.5   Production of Gluten Exorphins

The gluten exorphins are released from the high molecular weight glutenin and the gliadorphin is generated from alpha gliadorphin of wheat. The analysis of these opioid peptides is done through simulated gastrointestinal digestion (SGID) utilizing a combination of various enzymes. The purification of these peptides is done through analytical HPLC and the sequence is determined on a sequencer. More specifically, the researchers isolated wheat gluten from wheat, dissolved in water at a concentration of 50 mg/ml, and then digested with pepsin. The latter enzyme was added at a concentration of 0.5 mg/mg of gluten protein. The pH was adjusted to 2 with 0.02N HCl and incubated at 36 °C for 17 h. The reaction was stopped by increasing the pH to 7.0 with 1N NaOH. The solution was heated and then centrifuged. Lyophilization of the supernatant was then carried out. The digest previously hydrolyzed with pepsin was further digested with trypsin, chymotrypsin, or thermolysin at a concentration of 0.5 mg/ml. The digestion was carried out at 36 °C for 5 h. The reaction was then stopped by boiling and then centrifuged. The peptides were separated by reverse-phase high-performance liquid chromatography. The analytical HPLC had an octadecyl silane column. The sample (10 mg) was loaded on this analytical HPLC column. Gradient elution was performed in a linear pattern between 0 and 40% acetonitrile (ACN) containing trifluoroacetic acid (0.05%) at a rate of 10 mg/ml. The detection of the peptides was done with the UV-Visible detector at 230 nm. The fractions were collected and then evaporated on the centrifugal concentrator and the opioid activities of different fractions were performed as described in the previous section. The fractions with the opioid activity were further purified on an ODS, phenyl silica, cyanopropyl, and phenyl silica columns. The elution was performed through gradient elution with a linear gradient between 0 and 50% acetonitrile

(ACN) containing 0.05% TFA or 10 nM sodium-potassium phosphate buffer at the rate of 1 ml/min. The detection of the peptides was again done through a UV-Visible detector between 215 and 235 nm. The active fractions were once again analyzed on the ODS column. The sequence of the amino acids of the purified peptides was done with the protein sequencer (477A) (Applied Biosystems Inc.). The standard peptides used in the present investigation were synthesized through Sam 2 peptide synthesizer (Bioscxch Inc.). The peptide was deprotected by anisole or hydrogen fluoride method and purified by reverse-phase HPLC on an ODS column. The researchers found opioid activities in four fractions and were denoted as I, II, III, and IV. The latter fractions were eluted at 23, 25, 29, and 34 ACN, respectively. Moreover, these investigators further purified four active peptides on RP-HPLC with an ODS column and recovered four pure opioid peptide fractions. The sequence analysis revealed four sequences Gly-Tyr-Tyr-Pro-Thr, Gly-Tyr-Tyr-Pro, Tyr-Gly-Gly-Trp, and Tyr-Gly-Gly-Trp-Leu. These peptides were designated as gluten exorphins A5, A4, B4, and B5, respectively. Furthermore, the results indicated that exorphin A5 to be an interspersed repeated sequence that repeated itself 15 times in the primary structure of glutenin (Fukudome and Yoshikawa 1992). The same investigators again in 1993 detected novel wheat-derived exorphin named as gluten exorphin C. They followed the same protocol of enzymatic as earlier (Fukudome and Yoshikawa 1992). Moreover, the same methodology was employed for analysis of the peptides. In this trial, the novel gluten exorphin C whose sequence is Tyr-Pro-Ile-Ser-Leu was detected through the utilization of pepsin, trypsin, and chymotrypsin. The IC_{50} values assessed were 40 and 13.5 µM through GPI and MVD assays, respectively. This novel peptide has structure deviated compared to other exorphins and endorphins because it possessed only one tyrosine residue in the whole peptide at the amino-terminal (Fukudome and Yoshikawa 1993). Quite evidently this peptide was generated through only three gastrointestinal peptidases; it implies that there is enough probability that this exorphin is produced through gastrointestinal digestion in vivo. Studies have established that oral administration of peptic hydrolysate of wheat gluten affected the hormone release and gastrointestinal motility and the effects were revered on the administration of naloxone (Schusdziarra et al. 1981, 1983; Morley et al. 1983). Therefore it is highly probable that this novel peptide may be accountable for such regulation of biological functions.

Fukudome and coworkers in 1997 tried to investigate the presence of gluten exorphins from gluten hydrolysate. The digestion was performed with pepsin and thermolysin. The enzyme-linked immunosorbent assay (ELISA) revealed the presence of gluten A5 immunoreactive material (Gly-Tyr-Tyr-Pro-Thr) in this hydrolysate. Moreover, the investigators isolated the other gluten exorphins like A5, B5, and B4. It was, therefore, suggested that these food-derived opioid peptides may be released in the gastrointestinal tract. The amount of gluten exorphin A5 from pepsin-elastase digest was found larger compared to that of the pepsin-thermolysin digest. Furthermore, it was found that the sequence of gluten exorphin A5 is repeated 15 times in the primary structure of glutenin. The commercially synthesized peptides were used for the determination of the region from which gluten exorphin A5 is released (Fukudome et al. 1997).

In 2006, Fanciulli and coworkers exploited the more advanced technique of Liquid Chromatography-Mass Spectroscopy for analysis of wheat-derived opioid peptides in the gastrointestinal tract and cerebrospinal fluid (CSF). The sample (CSF) was loaded into the system coupled with reverse-phase HPLC with a C_{18} hydrophobic column. Gradient elution was performed with eluent A (water and 0.6% acetic acid) and eluent B (acetonitrile or methanol) (75:25, v/v). The LC-MS system was set in a way that the first 4 min solvent was diverted as waste. After this time, the remaining eluent was directed towards a mass spectrometer that showed the presence of the peptide in the sample. The investigators confirmed the presence of gluten Exorphin A5 in the CSF. This highly sophisticated technique presented evidence of the production of gluten derived opioid peptides in the gastrointestinal tract and their presence in the cerebrospinal fluid (Fanciulli et al. 2006). The same investigators extended this research work in the next year and used the same advanced technique (LC-MS) for the detection of gluten exorphin B5 in the CSF. The authors successfully confirmed the presence of this peptide in the CSF. Therefore, this study further validated the earlier results of the production and presence of the wheat-derived opioid peptides in the CSF (Fanciulli et al. 2007).

Karl and coworker in 2007 analyzed the presence of wheat gluten and gliadin derived opioid peptides in the urine samples of autistic patients. The urine samples were centrifuged and then filtered through 0.22 mm cellulose acetate filters. The filtered samples were applied on the reverse phase-HPLC and eluted with TFA and ACN. The detection of the peptides was performed with the UV-Visible detectors at 215, 280, 325 nm for peptide bonds, aromatic amino acids, and indole groups, respectively. The fractions corresponding to the peaks containing expected food-derived exorphins were collected and then lyophilized. The latter was reconstituted in methanol/water and formic acid (10 mM) and buffer (300 ml). The latter solution was stirred at room temperature and then loaded on mass spectrometer for sequence analysis. The investigators were able to detect glutenin derived gluten exorphins like A4, A5, B4, and B5. Moreover, the presence of gliadorphin-7 (Y-P-Q-P-Q-P-F) was also detected (Reichelt et al. 2012). These observations lead to the conclusion that these gluten exorphins are released in the gastrointestinal tract of autistic patients and then transported to the systemic circulation. Therefore, these findings present clues for the various physiological manifestations in these individuals through these exorphins.

6.6   Biological Role of Gluten Exorphins

6.6.1   Celiac Disease

The epitopes generated from the wheat gluten are among potential candidates for the development of celiac diseases (CD) in individuals that are genetically susceptible to this disease (Troncone and Jabri 2011). Moreover, the additional data further support the deleterious impact of the gluten diet on human health. Nevertheless, there is no mutual consensus in this perspective. This is because of the lack of gastrointestinal symptoms even in patients with established CD. The common symptoms associated with the CD include diarrhea, malabsorption, muscle wasting and weight loss, bloating, flatulence, and abdominal pain. More interestingly, some patients with established CD observed through assessment of duodenal biopsy and specific immunoglobins (Troncone and Jabri 2011; Strohle et al. 2013) still do not show classical symptoms. The latter condition is known as "asymptomatic CD" (ACD). The patients that suffer from ACD further demonstrate psoriasis (Pinto-Sanchez et al. 2015), sleep apnea in children (Parisi et al. 2015), diabetes mellitus type I (Bybrant et al. 2014; Hansson et al. 2015), neoplasia (Brito et al. 2014), severe hypoglycemia in diabetes mellitus type I (Khoury et al. 2014), depression (Pinto-Sanchez et al. 2015), atopic dermatitis (Ress et al. 2014), subclinical synovitis in children (Iagnocco et al. 2014), schizophrenia (Jackson et al. 2012), irritable bowel syndrome (IBS) (Pinto-Sanchez et al. 2015), and autism (Knivsberg et al. 2002). These all manifestations suggest that consumption of gluten is associated with the development of these complications. The gluten that is a complex of glutenin and gliadin after consumption is hydrolyzed in the gastrointestinal tract to release gluten exorphin and gliadorphins including gliadorphin-7. Already studies have reported the presence of A4, A5, B4, and B5. Moreover, the presence of gliadorphin-7 (Y-P-Q-P-Q-P-F) was also detected (Reichelt et al. 2012). Furthermore, other researchers have detected the presence of gluten exorphin A5 and B5 in cerebrospinal fluid (Fanciulli et al. 2006, 2007). The intactness of the proline-rich exorphin in the gut is dependent on the expression of the prolyl proteases like dipeptidyl peptidase (DPPIV). The uniqueness of these proteases is the specificity for the proline residues that are otherwise not common for other proteolytic enzymes. These proteases cleave these exorphins in dipeptides at the proline residues (Augustyns et al. 2005; De Meester et al. 2000; Vanhoof et al. 1995). Moreover, the resulting di- and tripeptide act as the inhibitors (competitive) of the aforementioned enzymes (DPPIV) (Augustyns et al. 1999; Mentlein 1999; Rahfeld et al. 1991). Studies have further demonstrated that the amino acids generated on complete digestion of gliadin provide obstacles for the presence of gluten epitopes that are otherwise responsible for the inducing proinflammatory responses in genetically susceptible individuals (Bethune and Khosla 2012; Hausch et al. 2002). A deficiency or the inactivity of DPPIV provides more chances for the production and existence of gluten exorphins in the gut (Hausch et al. 2002; Detel et al. 2007; Ozuna et al. 2015; Shan et al. 2002). Moreover, other observations show the gluten and milk-derived gliadin and casein, respectively, demonstrate more substrate specificity for prolyl proteases like DPPIV (Kikuchi et al. 1988; Castillo et al. 2012). Therefore, it presents a possible justification for the patients with established CD are asymptomatic. If/when the prolyl protease (DPPIV) is inhibited by gliadin results in the increased production of gliadin exorphins and may show symptoms like abdominal pain associated with CD.

The exact mechanism that leads to the establishment of ACD is still not clear and various signal cascade mechanism needs to be explored in this perspective. Nevertheless, Pruimboom and Punder in 2015 presented this aforesaid putative mechanism to illustrate the phenomenon. The authors finally concluded that incomplete hydrolysis of gluten results in gluten exorphins that may be responsible for the lack of conventional symptoms (gastrointestinal) of the subjects suffering from gluten consumption diseases. Furthermore, the incomplete digestion of the gluten causes inhibition of DPPIV and results in the manifestation of additional intestinal symptoms in these patients. If/when this putative model is proven correct, only then these patients should be recommended to take diets free from gluten to avoid gluten-induced complications (Pruimboom and de Punder 2015). Cade and coworkers tried to assess the Dohan's hypothesis that establishes a link between the absorption of gluten derived exorphins and increased risk for development of schizophrenia. These workers investigated a group of autistic children and exorphin absorption. In these neurological manifestations (autism and schizophrenia), a similar trend was observed in peptides when molecular screening was performed through Sephadex G-15. The results showed an increase in IgG against the gliadin proteins in 86% schizophrenic and 87%autisctic patients. Moreover, a high IgA titer was observed in 30% and 86% of children with autism and schizophrenia. Therefore, these results support the earlier hypothesis that autistic and schizophrenic patients have an increased tendency for the absorption of gluten derived opioid peptides. Therefore these peptides may have links for the manifestations of these health complications (Cade et al. 2000).

6.6.2 Prolactin Secretion

The wheat-derived opioid peptides as described in the previous section have been demonstrated to have a role in the secretion of prolactin. Literature in the early past has established the role of opioid peptides in the secretion of hormones from the pituitary gland in humans as well as in animal models. Fanciulli and coworkers in 2002 for the first time studied the role of gluten exorphin in pituitary function in rat models. Male rats were divided into four groups: intracerebroventricular (ICV) vehicle, gluten exorphin B5 (200 μg ICV), intraperitoneal administration of naloxone followed by vehicle ICV, and intraperitoneal administration of naloxone followed by gluten exorphin B5. The blood samples were collected for estimation of prolactin and growth hormone at intervals (90 min) after administration of exorphin or vehicle. The results indicated non-significant changes in the growth hormone secretion in this experiment. However, a significant change was observed in the secretion of prolactin through gluten exorphin B5 when administered through the ICV route. Therefore, the authors suggest a possible role of this exorphin in pituitary function and that too may occur through conventional opioid receptors (Fanciulli et al. 2002). The same investigators extended the research work on gluten exorphin B5 and the rats were divided into four groups. However, in this experiment, the route of the administration was changed. The animals were divided into four groups: intravenous vehicle, intravenous administration of gluten exorphinB5 given at a dose of 3 mg/kg body weight, intraperitoneal administration of naloxone followed by intravenous administration of vehicle, intraperitoneal administration of naloxone followed by intravenous administration of peptide. The blood samples were taken at intervals (60 min) after peptide or vehicle administration. The researchers found that the exorphin B5 when given at a dose of 3 mg/kg body weight significantly increased the prolactin levels, the effect was inhibited on naloxone administration. The authors concluded that the exorphin B5 has a role in the secretion of prolactin when given through peripheral role and the regulation occurs through conventional opioid pathways as stated earlier. However, the dose of the peptide given will have a possible role in the secretion of prolactin (Fanciulli et al. 2003). The same investigators extended this work in 2004 and in this experiment they studied the role of gluten exorphin B4 instead of B5 for the prolactin secretion in the rat model. The animals were divided into five groups against the four groups as in the previous trials. The groups were a vehicle, gluten exorphin B5 at a dose of 3 mg/kg body weight, gluten exorphin B4 at a dose of 3, 6, and 9 mg/kg body weight. The authors found that gluten exorphin B5 (dose of 3 mg/kg body weight) induced significant changes in the secretion of prolactin levels The authors found that this exorphin (B4) compared to earlier studied (B5) could not alter the secretion of prolactin in rats, although this peptide was given at doses three times higher compared to earlier analyzed peptide. The authors finally concluded that leucine at the carboxyl end is significant for its function in secreting prolactin (Fanciulli et al. 2004). The aforementioned studies establish that gluten exorphin has a role in the stimulation of prolactin levels in rat models. Moreover, this effect is abolished by the naloxone, an opioid receptor antagonist. The latter permits it transports through the blood–brain barrier (BBB) and in turn antagonizes the activity in the central and peripheral nervous system. To analyze the effects of this opioid peptide on prolactin levels occurring via inside or outside BBB, the same investigators pretreated animals with naloxone bromide instead of naloxone, the former does not cross the BBB. The results indicated that gluten exorphin-5 encouraged secretion of prolactin via opioid receptors that are present outside the BBB. It is well-established that exorphins are unable to secrete prolactin directly, however, the same occurs via reduced dopaminergic tone. Therefore these observations indicate that gluten exorphin 5 has a potential role in the modification of the brain neurotransmitter release without transporting through the BBB (Fanciulli et al. 2005).

6.6.3   Behavioral and Pharmacological Effects

Takahashi and coworkers in 2000 studied the role of gluten exorphin A5 in the pain modulation, observable behavior, and learning/memory processes in animal models (mice). The researchers found that this exorphin could not produce a pain-relieving effect or modulations in the morphine analgesia. Administration of this peptide intracerebroventricularly (ICV) was able to show statistically significant (although mild) pain management in a concentration-dependent fashion; however, it could not affect morphine analgesia. Oral administration of this peptide inhibited the antinociception prompted by socio-psychological stress. Intraperitoneal administration of this peptide eradicated both socio-psychological stress and stress-induced antinociception. Although intracerebroventricular administration of this peptide inhibited FS-SIA, nevertheless, it encouraged PSY-SIA. The exorphin administered through these different routes could not affect forced swim-SIA. Moreover, the oral administration of this peptide tends to increase the retention time on open arms observed through the elevated plus-maze test. Additionally, it was found that oral administration of this peptide during the post-training period time of learning or memory processes significantly elevated the dormancy into the dark element in the passive avoidance test. It indicates that this exorphin also expedites the attaining or consolidation phenomenon of learning or memory. Thus, gluten exorphin A5 has been found to produce various effects not only in the peripheral nervous systems but also in the central nervous system (Takahashi et al. 2000). Dohan has proposed that surplus concentration of wheat or casein derived exorphins may be the potential candidates for the establishment of neurological manifestations like that of schizophrenia (SZ) at the minimum in susceptible subjects. The observations generated in rodent models convey deep comprehensions for the credibility of Dohan's exorphin hypothesis. These animal models provide direct means to analyze the effects of such exorphins. Studies have established that exorphins like gliadorphin-7, β-casomorphins, rubiscolins, and soymorphins were psychologically lively in numerous models evaluating spontaneous behavior, nociception, and memory. Therefore, the food-derived exorphins are potential candidates for modulation of rodent behavior. However, more studies with explored biochemical mechanisms are needed to study these exorphins to reach final conclusions. Although it is well established that these exorphins bind the opioid receptors, it is further to be explored that these peptides are produced in sufficient quantities to show a biological, behavioral, or psychological effect. The limiting factor like absorption in the gastrointestinal tract and transport across the blood–brain barrier needs further deep research. Since, there are no direct correlations between the consumption of casomorphins and gluten exorphins and the development of schizophrenia. However, documented correlations between casein and gluten intake and schizophrenia sensitivity justify enhanced patient profiling and nutritional intervention where needed (Lister et al. 2015). Artemova and coworkers have suggested that food-derived exorphins have a potential role in the suppression of aggregation and encourage a recurrence of partially unfolded stress proteins. The commercially synthesized exorphins with sequences similar to gluten exorphin C and other opioid peptides like rubiscolin and hemorphin have shown protective effects in this perspective. These beneficial effects have been demonstrated through various spectroscopic methods including turbidity, fluorescence, dynamic light scattering, and circular dichroism. These researchers have found a dose-dependent effect (suppression) of light scattering intensity of aggregates of lactalbumin and alcohol dehydrogenase in the presence of these exorphins. The quantity of nanoparticles that show large radii (hydrodynamic) is shifted to lower ones, associated with prolonged time of aggregation. The exorphins in the refolding medium were depicted to help reactivation of carbonic anhydrase and increase the production of the carbonic anhydrase soluble protein. These findings, therefore, suggest that exorphins from food sources may be potential exogenous complements that may aid the endogenous defense mechanisms of humans under stress conditions (Artemova et al. 2010).

6.6.4 Hormonal and Gastrointestinal Functions

The effect of gluten exorphins was observed on the gastrointestinal transit time, hormone secretion, satiety, and small bowel mucosal integrity. It was found that pepsin-hydrolysate of the gluten increased the intestinal transit time, the same effect was found to get reverse on the administration of naloxone. Similarly, this gluten hydrolysate increased the secretion of somatostatin in plasma that was reversed on the naloxone administration. However, the results indicated that there is no significant effect on the secretion of serum cortisol and gastrin. Moreover, no changes were observed on the small bowel integrity on the administration of exorphins. Although many reports have established the role of endorphins in the modulation of appetite, nevertheless, these investigators could not observe any significant effect on the number of calories taken or satiety perception. Therefore, the results demonstrate a potential impact of these exorphins from various diets in the management of intestinal function (Morley et al. 1983).

The researchers have studied the effect on both digested and undigested gluten on postprandial glucagon and insulin secretion in conscious dogs. The dogs were administered with 25 g digested and undigested gluten. The results indicated that the former type of gluten (digested) showed a significant and more rapid increase in glucagon and insulin secretion compared to the latter type of gluten (undigested). It was found that the incremental insulin concentration was $104\pm$ μU/ml when digested gluten was administered compared to undigested one 58 ± 7 Mu/ml. The corresponding levels of glucagon were 426 ± 25 pg/ml in digested gluten compared to undigested gluten 302 ± 20 pg/ml. Moreover, it was found that the co-administration of naloxone (4 mg, intragastric) decreased the induced levels and increased the glucagon levels; however, non-significant changes were observed from undigested gluten. When the naloxone was given intravenously no changes in the levels of hormone (insulin and glucagon) secretion were observed. In conclusion, these findings demonstrate that gluten hydrolysate contains some opioid exorphins that stimulate the postprandial glucagon and insulin secretion (Schusdziarra et al. 1981).

References

Aldrich JV, McLaughlin JP (2009) Peptide kappa opioid receptor ligands: potential for drug development. AAPS J 11(2):312–322

Artemova NV, Bumagina ZM, Kasakov et al (2010) Opioid peptides derived from food proteins suppress aggregation and promote reactivation of partly unfolded stressed proteins. Peptides 31 (2):332–338

Augustyns K, Bal G, Thonus et al (1999) The unique properties of dipeptidyl-peptidase IV (DPP IV / CD26) and the therapeutic potential of DPP IV inhibitors. Curr Med Chem 6(4):311–327

Augustyns K, Van der Veken P, Senten K et al (2005) The therapeutic potential of inhibitors of dipeptidyl peptidase IV (DPP IV) and related proline-specific dipeptidyl aminopeptidases. Curr Med Chem 12(8):971–998

Bethune MT, Khosla C (2012) Oral enzyme therapy for celiac sprue. Methods Enzymol 502:241–271

Brito MD, Martins A, Henrique R, Mariz J (2014) Enteropathy-associated T cell lymphoma as a complication of silent celiac disease. Hematol Rep 6(4):5612. https://doi.org/10.4081/hr.2014.5612

Bybrant MC, Ortqvist E, Lantz S et al (2014) High prevalence of celiac disease in Swedish children and adolescents with type 1 diabetes and the relation to the Swedish epidemic of celiac disease: a cohort study. Scand J Gastroenterol 49(1):52–58

Cade R, Privette M, Fregly M et al (2000) Autism and schizophrenia: intestinal disorders. Nutr Neurosci 3(1):57–72

Castillo GM, Reichstetter S, Bolotin EM (2012) Extending residence time and stability of peptides by protected graft copolymer (PGC) excipient: GLP-1 example. Pharm Res 29(1):306–318

Christison GW, Ivany K (2006) Elimination diets in autism spectrum disorders: any wheat amidst the chaff? J Dev Behav Pediatr 27(2):S162

De Meester I, Durinx C, Bal G et al (2000) Natural substrates of dipeptidyl peptidase IV. Adv Exp Med Biol 477:67–87

Detel D, Persic M, Varljen J (2007) Serum and intestinal dipeptidyl peptidase IV (DPP IV/CD26) activity in children with celiac disease. J Pediatr Gastroenterol Nutr 45(1):65–70

Fanciulli G, Dettori A, Tomasi PA et al (2002) Prolactin and growth hormone response to intracerebroventricular administration of the food opioid peptide gluten exorphin B5 in rats. Life Sci 71(20):2383–2390

Fanciulli G, Dettori A, Fenude E et al (2003) Intravenous administration of the food-derived opioid peptide gluten exorphin B5 stimulates prolactin secretion in rats. Pharmacol Res 47(1):53–58

Fanciulli G, Dettori A, Demontis MP et al (2004) Gluten exorphin B5 stimulates prolactin secretion through opioid receptors located outside the blood-brain barrier. Life Sci 76(15):1713–1719

Fanciulli G, Dettori A, Demontis MP et al (2005) Serum prolactin levels after administration of the alimentary opioid peptide gluten exorphin B4 in male rats. Nutr Neurosci 7(1):53–55

Fanciulli G, Azara E, Wood TD (2006) Quantification of Gluten Exorphin A5 in cerebrospinal fluid by liquid chromatography-mass spectrometry. J Chromatogr B Analyt Technol Biomed Life Sci 833(2):204–209

Fanciulli G, Azara E, Wood TD (2007) Liquid chromatography-mass spectrometry assay for quantification of Gluten Exorphin B5 in cerebrospinal fluid. J Chromatogr B Analyt Technol Biomed Life Sci 852(1-2):485–490

Fukudome S, Yoshikawa M (1992) Opioid peptides derived from wheat gluten: their isolation and characterization. FEBS Lett 296(1):107–111

Fukudome S, Yoshikawa M (1993) Gluten exorphin C. A novel opioid peptide derived from wheat gluten. FEBS Lett 316(1):17–19

Fukudome S, Jinsmaa Y, Matsukawa T (1997) Release of opioid peptides, gluten exorphins by the action of pancreatic elastase. FEBS Lett 412(3):475–479

Hansson T, Dahlbom I, Tuvemo T et al (2015) Silent coeliac disease is over-represented in children with type 1 diabetes and their siblings. Acta Paediatr 104(2):185–191

Hausch F, Shan L, Santiago NA et al (2002) Intestinal digestive resistance of immunodominant gliadin peptides. Am J Physiol Gastrointest Liver Physiol 283(4):G996–G1003

Hughes J, Smith TW, Kosterlitz HW et al (1975) Identification of two related pentapeptides from the brain with potent opiate agonist activity. Nature 258(5536):577–580

Iagnocco A, Ceccarelli F, Mennini M et al (2014) Subclinical synovitis detected by ultrasound in children affected by coeliac disease: a frequent manifestation improved by a gluten-free diet. Clin Exp Rheumatol 32(1):137–142

Jackson J, Eaton W, Cascella N et al (2012) A gluten-free diet in people with schizophrenia and anti-tissue transglutaminase or anti-gliadin antibodies. Schizophr Res 140(1–3):262–363

Janecka A, Fichna J, Janecki T (2004) Opioid receptors and their ligands. Curr Top Med Chem 4:1–17

Khoury N, Semenkovich K, Arbelaez AM (2014) Coeliac disease presenting as severe hypoglycemia in youth with type 1 diabetes. Diabet Med 31(12):e33–e36. https://doi.org/10.1111/dme.12488

Kikuchi M, Fukuyama K, Epstein WL (1988) Soluble dipeptidyl peptidase IV from terminal differentiated rat epidermal cells: purification and its activity on synthetic and natural peptides. Arch Biochem Biophys 266(2):369–376

Kitts DD, Weiler K (2003) Bioactive proteins and peptides from food sources. Applications of bioprocesses used in isolation and recovery. Curr Pharm Des 9:1309–1323

Knivsberg AM, Reichelt KL et al (2002) A randomized, controlled study of dietary intervention in autistic syndromes. Nutr Neurosci 5(4):251–261

Lister J, Fletcher PJ, Nobrega JN et al (2015) Behavioral effects of food-derived opioid-like peptides in rodents: implications for schizophrenia? Pharmacol Biochem Behav 134:70–78

Liu Z, Udenigwe CC (2019) Role of food-derived opioid peptides in the central nervous and gastrointestinal systems. J Food Biochem 43(1):e12629. https://doi.org/10.1111/jfbc.12629

Loukas S, Varoucha D, Zioudrou C et al (1983) Opioid activities and structures of alpha-casein-derived exorphins. Biochemistry 22:4567–4573

Mentlein R (1999) Dipeptidyl-peptidase IV, (CD26)–role in the inactivation of regulatory peptides. Regul Pept 85(1):9–24

Morley JE, Levine AS, Yamada T et al (1983) Effect of exorphins on gastrointestinal function, hormonal release, and appetite. Gastroenterology 84(6):1517–1523

Ozuna CV, Iehisa JC, Gimenez MJ (2015) Diversification of the celiac disease alpha-gliadin complex in wheat: a 33-mer peptide with six overlapping epitopes, evolved following polyploidization. Plant J 82(5):794–805

Parisi P, Pietropaoli N, Ferretti A et al (2015) Role of the gluten-free diet on neurological-EEG findings and sleep disordered breathing in children with celiac disease. Seizure 25:181–183

Pert CB, Snyder SH (1973) Opiate receptor: demonstration in nervous tissue. Science 179 (4077):1011–1014

Pert CB, Pasternak G, Snyder SH (1973) Opiate agonists and antagonists discriminated by receptor binding in brain. Science 182(4119):1359–1361

Pinto-Sanchez MI, Bercik P, Verdu EF et al (2015) Extra intestinal manifestations of celiac disease. Dig Dis 33(2):147–154

Pruimboom L, de Punder K (2015) The opioid effects of gluten exorphins: asymptomatic celiac disease. J Health Popul Nutr 33:24

Rahfeld J, Schierhorn M, Hartrodt B et al (1991) Are diprotin A (Ile-Pro-Ile) and diprotin B (Val-Pro-Leu) inhibitors or substrates of dipeptidyl peptidase IV? Biochim Biophys Acta 1076(2):314–316

Reichelt KL, Tveiten D, Knivsberg AM et al (2012) Peptides' role in autism with emphasis on exorphins. Microb Ecol Health Dis 23:18958. https://doi.org/10.3402/mehd.v23i0.18958

Ress K, Annus T, Putnik U et al (2014) Celiac disease in children with atopic dermatitis. Pediatr Dermatol 31(4):483–488

Satoh M, Minami M (1995) Molecular pharmacology of the opioid receptors. Pharm Ther 68:343–364

Schusdziarra V, Henrichs I, Holland A (1981) Evidence for an effect of exorphins on plasma insulin and glucagon levels in dogs. Diabetes 4:362–364

Schusdziarra V, Rewes B, Lenz N (1983) Evidence for a role of endogenous opiates in postprandial somatostatin release. Regul Pept 6(4):355–361

Shan L, Molberg O, Parrot I et al (2002) Structural basis for gluten intolerance in celiac sprue. Science 297(5590):2275–2279

Sharma SK, Klee WA, Nirenberg M (1975) Dual regulation of adenylate cyclase accounts for narcotic dependence and tolerance. Proc Natl Acad Sci 72(8):3092–3096

Strohle A, Wolters M, Hahn A (2013) Celiac disease-the chameleon among the food intolerances. Med Monatsschr Pharm 36(10):369–380

Takahashi M, Fukunaga H, Kaneto H et al (2000) Behavioral and pharmacological studies on gluten exorphin A5, a newly isolated bioactive food protein fragment, in mice. Jpn J Pharmacol 84 (3):259–265

Tatham AS, Shewry PR (2008) Allergens to wheat and related cereals. Clin Exp Allergy 38 (11):1712–1726

Teschemacher H (2003) Opioid receptor ligands derived from food proteins. Curr Pharm Des 9:1331–1344

Troncone R, Jabri B (2011) Coeliac disease and gluten sensitivity. J Intern Med 269(6):582–590

Vanhoof G, Goossens F, De Meester I et al (1995) Proline motifs in peptides and their biological processing. FASEB J 9(9):736–744

Yamamoto N, Ejiri M, Mizuno S (2003) Biogenic peptides and their potential use. Curr Pharm Des 9(16):1345–1355

Yoshikawa M, Takahashi M, Yang S (2003) Delta opioid peptides derived from plant proteins. Curr Pharm Des 9(16):1325–1330

Zioudrou C, Streaty RA, Klee WA (1979) Opioid peptides derived from food proteins. J Biol Chem 254(7):2446–2449

Rubiscolins, Soymorphins, and Oryzatensin 7

Abstract

Rubiscolins, soymorphins, and oryzatensin are spinach RuBisCo, soy β-conglycinin, and rice albumin-derived opioid peptides, respectively. Rubiscolin-5 (YPLDL) and 6 (YPLDLF) have tyrosine at the N-terminal followed by a proline and leucine. Soymorphin-5 (YPFVV), 6 (YPFVVN), and 7 (YPFVVNA) show tyrosine at the N-terminal followed by proline and phenylalanine. Oryzatensin (GYPMYPLPR) shows a unique structure with glycine at the N-terminal followed by the tyrosine residue. Rubiscolins and soymorphins both act as opioid agonists for δ- and μ-opioid receptors, respectively. However, oryzatensin demonstrates antagonistic activity through μ-receptors. Rubiscolins enhance glucose uptake possibly through AMPK activation to increase GLUT4 translocation. Rubiscolin-6 stimulates food intake in the aged mice with an intact ghrelin resistance. These exorphins decrease the consumption of a high-fat diet possibly through activation of the central δ-opioid receptor system. These peptides are the potential exogenous complements that may aid the endogenous defense mechanisms of humans under stress conditions. Rubiscolins have a possible role in the regulation of the protective behavior in insects and the adaptation of young mammals. These exorphins have a role in anxiolysis and memory consolidation in mice. Soymorphins also play a role in anxiolysis, analgesia, food intake, and intestinal transit. Oryzatensin promotes phagocytic activity in PMNLs and the generation of superoxide ions from peripheral leukocytes. It has a potential role in immunomodulation through sensitization of cancer cells. Also, it improves amnesia induced by scopolamine and ischemia. Moreover, this peptide suppresses the food intake possibly through PGE2 generation and EP4 receptor activation.

7.1   Introduction

Bioactive peptides (BAP) are the fragments of proteins usually composed of 4–20 amino acids. Although, some of these BAPs exist in free forms, however, most of these functional fragments are released from the intact proteins. These are inactive within the whole proteins, however, they show activities when released from these large biomolecules. These are obtained from food sources like wheat, milk, soybean, rice, spinach, etc. The advanced bioanalytical techniques like chemical synthesis and recombinant DNA technique have enabled us to synthesis these peptides either from individual amino acid or from nucleic acids, respectively. There are numerous reports available that have demonstrated a positive impact of these BAPs on physiological processes in humans. However, during the last few decades, enough data has been generated that correlate the consumption of these peptides with various human illnesses like type 1 diabetes, cardiovascular diseases, sudden infant death syndrome, and others. Among the beneficial effects include the mineral binding, angiotensin-converting enzyme (ACE) inhibitory activity, antioxidative and antimicrobial activities. In this positive perspective, there is a growing interest for the scientific community, health professionals, food and pharmaceutical industry to devise methods for the introduction of a functional bioactive peptide derived food additives and nutraceuticals (Sánchez and Vázquez 2017).

The composition and sequence of amino acids in a BAP determines its biological activity, as these are generated from their parent protein molecules. When these BAPs interact with the endogenous receptors, they have the potential to alter various physiological processes (Fields et al. 2009). These BAPs may be endogenous or exogenous. The former is synthesized from the glands directly and the latter is obtained from the exogenous food sources. The exogenous BAPs that demonstrate the opioid activity is denoted as exorphins. These exorphins are released from animal sources like milk (β-casomorphins, lactorphins, casoxins, and lactoferroxins) (Bhat et al. 2015a, b). However, the plant proteins are also potential sources of BAPs like wheat (gluten exorphins, gliadorphins), soybean (soymorphins), spinach (rubiscolins), and rice (oryzatensin) (Carrasco-Castilla et al. 2012; Lee and Hur 2017). The association between the structure and function of various BAPs is not well-established to date. Nevertheless, there is some regularity in these BAPs from diverse food sources. For example, most of these BAPs are composed of two to twenty amino acids (Möller et al. 2008), presence of tyrosine at the amino-terminal, presence of proline, arginine, or lysine and some hydrophobic amino acids. Moreover, all exorphins have selectivity towards the mu-opioid receptor (MOP), delta-opioid receptor (DOP), and kappa-opioid receptor (KOP). Furthermore, they are resistant to the digestion of most of the proteolytic enzymes (pepsin, trypsin, and chymotrypsin) due to an abundance of proline. However, most of these exorphins are susceptible to proteolytic cleavage of prolyl proteases like dipeptidyl peptidase or brush border enzymes (Kitts and Weiler 2003). The food-derived exorphins like soymorphins, rubiscolins, and oryzatensin is the topic of the discussion for this chapter.

7.2 Structure

7.2.1 Rubiscolins

These exorphins are derived from spinach RuBisCo. The general structure of the rubiscolin consists of tyrosine at the amino-terminal followed by a proline and leucine. Rubiscolin-5 has an amino acid sequence of YPLDL and rubiscolin-6 as YPLDLF.

7.2.2 Soymorphins

These exorphins are derived from soy β-conglycinin. The general structure of the soymorphins consists of tyrosine at the amino-terminal followed by proline and phenylalanine. All these amino acids are essential for the determination of selectivity for the opioid receptors. Soymorphin-5 has an amino acid sequence of YPFVV, soymorphin-6 with a sequence of YPFVVN, and soymorphin-7 has an amino acid sequence of YPFVVNA.

7.2.3 Oryzatensin

This exorphin is derived from rice soluble protein (albumin) and is composed of nine amino acid residues. The general structure of the oryzatensin consists of GYPMYPLPR. It has a unique structure with glycine at the amino-terminal followed by the tyrosine residue quite similar to gluten exorphin A4 and A5. The structure of these exorphins, i.e., rubiscolins, soymorphins, and oryzatensin, is shown in Fig. 7.1.

7.3 Classification

7.3.1 Rubiscolins 5 and 6

These food-derived opioid peptides also known as exorphins were first discovered in 2001. Yang and coworkers isolated these peptides from pepsin the hydrolysates of spinach protein (D-ribulose-1, 5-bisphosphate carboxylase/oxygenase (RuBisCo) (Yang et al. 2001). The latter is the most abundant protein of the plant kingdom and is ubiquitous. There are two types of rubiscolins denoted as rubiscolin-5 and rubiscolin-6. The former has a sequence Tyr-Pro-Leu-Asp-Leu) and the latter has an amino acid sequence Tyr-Pro-Leu-Asp-Leu-Phe. These opioid peptides have selectivity towards the δ-opioid receptor observed through [^{3}H] deltorphin II-binding assay. These also demonstrate opioid activity observed through mouse vas deferens assay (Yang et al. 2001). The larger opioid peptide (rubiscolin-6) shows a higher affinity for the opioid receptor. The sequence of rubiscolin-6 is found to be conserved and is located in the large subunit of RuBisCo. Reports have shown

Fig. 7.1 Structures of rubiscolins, soymorphins, and rubiscolin

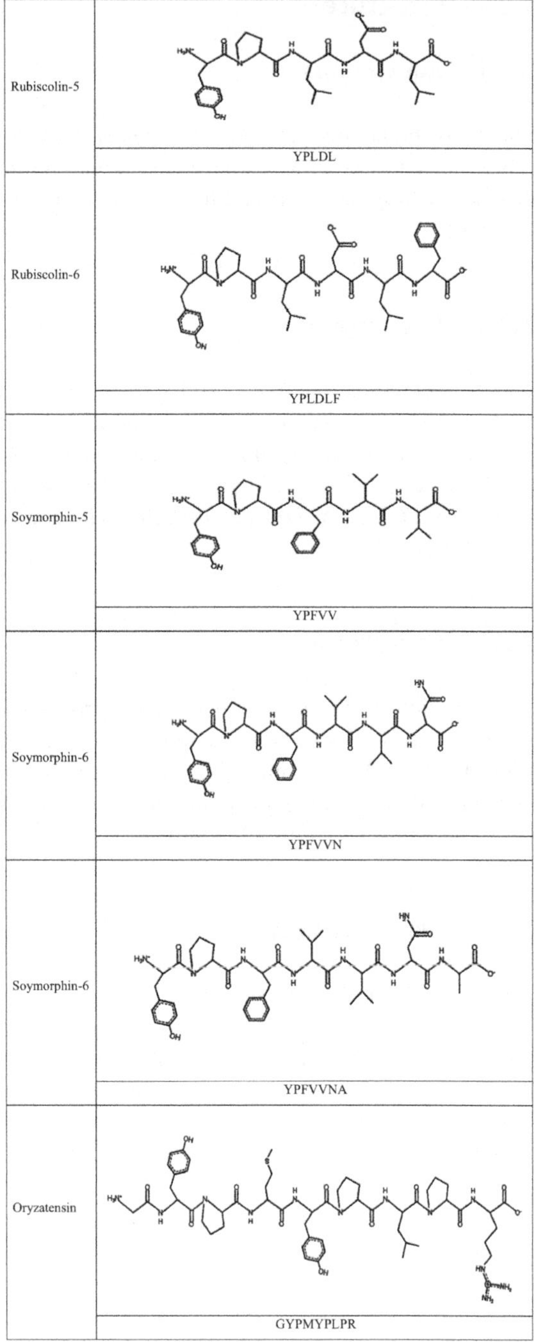

that when leucine is substituted at position 3rd in the rubiscolin with isoleucine, it increased the δ opioid activity (Yang et al. 2003). This substituted opioid peptide does naturally occur in some algae (Euglena gracilis, Euglena stellata, and Mesostigma viride). Therefore, it suggests that the peptide that will be isolated from these algae-based foods may be more potent δ-receptor ligands. Rubiscolin-5 has a molecular formula of $C_{30}H_{45}N_5O_9$ and molecular mass 619.7 g/mol. Rubiscolin-6 has a molecular formula of $C_{39}H_{54}N_6O_{10}$ and molecular mass 766.9 g/mol.

7.3.2 Soymorphins 5, 6, and 7

The soymorphins are usually chemically synthesized peptides that have a sequence similar to the peptide Tyr-Pro-Phe-Val that is derived from the β-conglycinin β-subunit of soy. The sequence analysis shows that this sequence is similar to human βh-casomorphin-4. Soymorphin-5 has a sequence Tyr-Pro-Phe-Val-Val, soymorphin-6 Tyr-Pro-Phe-Val-Val-Asn, and soymorphin-7 Tyr-Pro-Phe-Val-Val-Asn-Ala. Soymorphins opioid peptide has selectivity towards the μ-opioid receptor. Among these three soymorphins, the soymorphins-5 demonstrates the highest opioid activity (Ohinata et al. 2007). Symorphin-5 has molecular mass 623.3308 g/mol, soymorphin-6 has 737.3736 g/mol, and soymorphin-7 with a molecular mass of 808.4106 g/mol. More research is needed to analyze the production of these peptides in large quantities and their opioid activity.

7.3.3 Oryzatensin

Researchers have isolated a novel peptide from the hydrolysate of the soluble protein of rice (albumin). The hydrolysis was performed with the trypsin and has a sequence Gly-Tyr-Pro-Met-Tyr-Pro-Leu-Pro-Arg. It was demonstrated to have anti-opioid activity when observed through GPI assay. This peptide depicted biphasic contractions on ileum that showed rapid followed by slower contraction. When inhibitors like atropine and tetrodotoxin were used in the assay, it was observed that slower contractions were mediated by the cholinergic nervous system. The results showed that this peptide has a weak affinity towards μ-receptors and linked with slower contraction. This peptide was found to demonstrate phagocytosis-promoting activity in polymorphonuclear leukocytes derived from humans. This rice derived exorphin was found to amplify the generation of superoxide anion in vitro in peripheral leukocytes (human) (Takahashi et al. 1994). The classification of rubiscolins, soymorphins, and oryzatensin is given in Table 7.1.

Table 7.1 Classification of rubiscolins, soymorphins, and oryzatensin (table modified from Garg et al. 2016)

Protein source	Opioid peptide	Sequence	ORS	A/ AN	Mr (g/mol)	MF
Spinach RuBisCo	Rubiscolin-5	YPLDL	δ	A	619.7	$C_{30}H_{45}N_5O_9$
	Rubiscolin-6	YPLDLF	δ	A	766.9	$C_{39}H_{54}N_6O_{10}$
Soy β-conglycinin	Soymorphins-5	YPFVV	μ	A	623.3	$C_{33}H_{45}N_5O_7$
	Soymorphins-6	YPFVVN	μ	A	737.3	$C_{37}H_{51}N_7O_9$
	Soymorphins-7	YPFVVNA	μ	A	808.4	$C_{40}H_{56}N_8O_{10}$
Rice soluble protein	Oryzatensin	GYPMYPLPR	μ	AN	1093.3	$C_{52}H_{76}N_{12}O_{12}S$

ORS opioid receptor selectivity, *A* agonist, *AN* antagonist, *Mr* molecular weight, *MF* molecular formulae

7.4   Opioid Activity

7.4.1   Rubiscolin

The expression of opioid activity in rubiscolin 5 and 6 is attributed to amino-terminal tyrosine and proline. The measured IC_{50} values for rubiscolin 5 and 6 are 2.09 and 0.093 μM, respectively. The same has been measured through receptor (δ) binding assay. These are selective for δ opioid receptors and demonstrate antinociceptive activity (Yang et al. 2001). Rubiscolin-6 augments memory development (Yang et al. 2003) shows anxiolytic activity (Hirata et al. 2007), orexigenic activity (Kaneko et al. 2012a, b; Miyazaki et al. 2014), and decreases intake of high fat (Kaneko et al. 2014).

7.4.2   Soymorphins

Soymorphins are chemically synthesized peptides sharing the sequences of β-conglycinin of spinach. The two exorphins soymorphins-5 and -7 demonstrated lower IC_{50} values through GPI compared to MVD assay (Ohinata et al. 2007). They have selectivity towards the μ-opioid receptors and demonstrate anxiolytic properties (Ohinata et al. 2007). Soymorphin-5 decreases food intake and prolongs the small intestine transit (Kaneko et al. 2010). It has a role in metabolism and enhances the β-oxidation and in turn energy expenditure (Yamada et al. 2012). The analogs chemically synthesized through amidation of soymorphin-5 encouraged movement and suppressed anxiety in juvenile rats (Chesnokova et al. 2014). These aforesaid properties remark potential of soymorphin-5 in controlling human complications like obesity, diabetes, and anxiety (Kaneko et al. 2010; Yamada et al. 2012).

7.4.3 Oryzatensin

Oryzatensin was purified from rice soluble protein trypsin hydrolysate. It acts as an opioid antagonist with MOP selectivity. This peptide depicted biphasic contractions on ileum that showed rapid followed by slower contraction. The use of inhibitors (atropine and tetrodotoxin) in the opioid activity determining assays demonstrated slower contractions were mediated by the cholinergic nervous system. This peptide exhibited a weak affinity towards μ-receptors associated with slower contraction. It shows phagocytosis-promoting activity in polymorphonuclear leukocytes derived from humans. This rice derived exorphin was found to amplify the generation of superoxide anion in vitro in peripheral leukocytes (human) (Takahashi et al. 1994). The opioid activity of rubiscolins, soymorphins, and oryzatensin is shown in Table 7.2.

7.5 Production

7.5.1 Rubiscolins

Ribulose-1, 5-bisphosphate carboxylase/oxygenase (RuBisCo) protein from spinach (50 mg) was dissolved in 5 ml distilled water. The pH was adjusted to 2.0 with 0.1 N HCl and pepsin ($E/S = 1/50$, w/w) was added to this protein solution. The reaction (pepsin hydrolysis) was carried out for 4 h at 37 °C. After completion of the reaction, the same was stopped by an increase in pH to 8.0 on the addition of 1N NaOH. The digest was divided into two fractions, one of the fractions was further digested with leucine aminopeptidase ($E/S = 1/50$, w/w) for 8 h at 37 °C. This was followed by boiling to stop the reaction. The peptides in the hydrolysate were analyzed by RP-HPLC fitted with the ODS column. Gradient elution was done with acetonitrile (0–50%/50 min) with added trifluoroacetic acid (0.1% 10 ml/min) and dried with a

Table 7.2 Opioid activity of rubiscolins, soymorphins, and oryzatensin (table modified from Garg et al. (2016))

Exorphins	Opioid activity (IC50 μM)		μ/δ ratio	Reference
	GPI	MVD		
Rubiscolin-5	1110.0	51.000	21.8	Yang et al. (2001)
Rubiscolin-6	748.00	24.400	30.7	
Soymorphin-5	6.0000	50.000	0.12	Ohinata et al. (2007)
Soymorphin-6	9.2000	32.000	0.28	
Soymorphin-7	13.000	50.000	0.26	
Oryzatensin	5.0000			Takahashi et al. (1994)

centrifugal concentrator. On the other hand commercially synthesized standard peptides with sequence, YPLDLF and YPLDL were run the same column and their retention times were noted. This was followed by the collection of fractions from hydrolysates corresponding to the standard peptides. The fractions so obtained were again purified on the phenethyl silica column. Gradient elution as discussed above was performed for purifying fractions on the cyanopropyl silica column. The fractions were freeze lyophilized on the centrifugal concentrator. The sequence of amino acids in the peptides was done with Protein Sequencer or LC-MS. A precursor peptide of 14 amino acids was synthesized that was further digested (2 mg/ml) with pepsin, elastase, and chymotrypsin ($E/S = 1/20$) for 2 h at 37 °C. It was divided into two digests, one was further digested with leucine aminopeptidase ($E/S = 1/20$) for 1 h at 37 °C. The digestion with pepsin was stopped by increasing pH through the addition of NaOH. After completion of reactions, these were stopped by boiling. The presence and the number of peptides released in the hydrolysate were calculated from the peak area on RP-HPLC corresponding to the standard peptides at 280 nm with the help of UV-Visible detector. The results indicated that rubiscolin-5 was produced from spinach RuBisCo through pepsin hydrolysis (0.17% mol/mol RuBisCo large subunit). The addition of LAP further increased the yield of this exorphin to 5.0%. However, these conditions could not generate rubiscolin-6. The synthesized 14 amino acid residue precursor corresponding to 98 to 111 of spinach RuBisCo large subunit after pepsin hydrolysis released both rubiscolin-5 (YPLDL, 0.30%) and rubiscolin-6 (YPLDLF, 0.04%) (Yang et al. 2001).

7.5.2　Soymorphins

Ohinata and coworker analyzed the release of soymorphins (5, 6 and 7) from the soy β-conglycinin fraction. Additionally, they checked the release of these exorphins from a chemically synthesized fragment of β-subunit of β-conglycinin from 318 to 333. They used pancreatic elastase and leucine aminopeptidase (LAP) for the hydrolysis protocols. For this hydrolysis purpose, they used 0.1 and 10 mg/ml of synthetic peptide and β-conglycinin, respectively, with elastase (E/S: 1/20 4 h) and LAP (E/S: 1/7 2 h) at pH 7.5 and 25 °C. The authors used RP-HPLC for the analysis of the peptides and the sequence of the amino acids in the peptides was determined by protein sequencer. Gradient elution was performed with ACN in water containing formic acid (0.1%) and TFA (0.01%) from 5 to 95% (v/v) for 8 min and a flow rate of 0.2 ml/min on octadecyl column (ODS). The exorphins were detected using UV-Visible detectors and electrospray MS in positive ion mode. The amount (yield) of soymorphins was calculated on the peak area on RP-HPLC. The results indicated that soymorphin-5 is released from the synthetic peptide and soy β-conglycinin with 36.5 and 9.1 mol %, respectively. Moreover, it was found that soymorphins-6 and -7 are not produced from either of the two aforesaid sources (Ohinata et al. 2007).

7.5.3 Oryzatensin

Takahashi and coworkers isolated oryzatensin opioid peptide from rice seeds *(Oryza sativa L. var. japonica cv. Nipponbare)*. These seeds were ground with an electric grinder. The rice powder so obtained (20 g) was homogenized with sodium chloride (1M, 140 ml) for 5 min. The same solution was stirred for 1 h. Centrifugation was then performed at 4000 rpm for half an hour. The precipitate was then taken and suspended in a sodium chloride solution and suspended for 1 h and then centrifuged. The first and second supernatants were mixed and then dialysis was performed against 0.1M sodium chloride. The pH was adjusted to 7.4 and the fraction was hydrolyzed with trypsin ($E/S = 1/100$) at 37 °C for 5 h. The reaction was stopped by boiling for 15 min and then stored at -20 °C until assay. The analysis of peptides was done on reverse phase-high performance liquid chromatography (RP-HPLC) using the ODS column. The sample (21 mg hydrolysate) was loaded on column and then the linear gradient of acetonitrile at a rate of 1%/min and 0.1% trifluoroacetic acid at a rate of 10 ml/min. Different fractions were collected on RP-HPLC and lyophilized on the centrifugal concentrator. The different fractions were determined for opioid activity on guinea-pig ileum. The fractions with opioid activity were further purified on the phenethyl silica column in presence of trifluoroacetic acid (0.1%) followed by purification on the ODS column in presence of phosphate buffer (10 mM, pH 7.4). Elution was performed by a linear gradient with 1% acetonitrile at 1 ml/min flow rate. The analysis of the amino acid sequence and peptide synthesis was performed as per Fukudome and Yoshikawa (1992). However, the peptide synthesis was performed as per the Fmoc method. The GPI assay demonstrated the opioid antagonistic activity of the fractions of tryptic hydrolysates of rice soluble protein. A novel peptide named oryzatensin was isolated with a sequence of GYPMYPLPR (Takahashi et al. 1994).

7.6 Physiological Role

7.6.1 Rubiscolins

As aforementioned in previous sections, these peptides are opioid agonists with selectivity towards δ-opioid receptors. It has further been established that opioid receptors on activation through their corresponding ligands have potential in increasing glucose uptake in cells (skeletal muscles) without the engagement of insulin. As previously discussed that rubiscolins are δ-opioid receptor ligands, therefore, may furnish a role in glucose metabolism in these cells. Studies have reported an increased protein expression of glucose transporter 4 (GLUT4) in addition to the protein kinase (AMPK) expression in L6 cells (rat skeletal muscle) and C2C12 cell lines (mouse myoblast). Also, studies have demonstrated augmented glucose (concentration-dependent) uptake in vivo in diabetic rats. For validation, the related opioid antagonists and inhibitors have blocked the effects mediated by these food exorphins. Therefore this opioid peptide suppressed the blood glucose and enhanced the GLUT4 protein expression in diabetic rats. Overall, it may be concluded that this peptide enhanced glucose uptake possibly through AMPK activation to increase GLUT4 translocation after rubiscolin-opioid receptor binding (Kairupan et al. 2019). More recently in 2019, Mitsumoto and coworkers have established that administration of rubiscolin-6 decreased the immobility time in restraint-stressed animals (mice) observed through tail suspension test, however, no changes were observed in the locomotor activity. More findings have shown that the opioid antagonist like naltrindole blocked the antidepressant-like activity induced earlier by rubiscolin-6. Therefore, it is suggested that this peptide induced antidepressant-like activity mediated through δ-opioid receptor activation (Mitsumoto et al. 2019). There is a significant role of β-arrestins in G protein-coupled receptor (GPCR) signaling and turn its manifestation through diverse physiological activities. Many reports have highlighted the adverse impact associated with natural and synthetic opioids. These hostile effects include simpler constipation to very complicated respiratory depression. It is assumed that the excessive recruitment of β-arrestins with GPCRs may be a potential risk factor. In this regard, there is a need for exploration of some alternate opioid receptor ligands that may activate these receptors through very minimal recruitment of β-arrestins. In this perspective, more recently in 2019, the researchers have revealed the rubiscolin-6 to activate GPCRs at δ opioid receptors. The advantage associated with these RuBisCo-derived peptides is the minimal recruitment of β-arrestins 1 and 2. Although the disadvantage of rubiscolins is the lower potency compared to the natural ligands deltorphin II and leu-enkephalin. Nevertheless, their significant distinction is the oral bioavailability and role in the various physiological processes discussed above. For the first time, these authors revealed the unique property of this exorphin in terms of minimal recruitment of β-arrestins at GCPRs (Cassell et al. 2019). The aforesaid unique attribute in addition to the oral bioavailability of this exorphin and activation at the δ-opioid receptors regards it as a potential drug against alcohol use disorder (Chiang et al. 2016; Robins et al. 2018) and an analgesic with lesser side effects. Previously, studies have established that aging results in a progressive decline in food intake, and consequently food-derived exorphins may have a role in reversing this effect. In this perspective adult (15 months) and old age (27 months) mice (C57BL/6N) were selected for the study. The body weight and the food intake in the aged mice were reduced compared to the adult mice. Moreover, other parameters including mesenteric and epididymis fat mass, triglyceride, leptin, and blood glucose levels were also suppressed in the aged mice. Moreover, the expression of agouti-related protein and neuropeptide was increased in aged mice. The investigators found that oral administration of rubiscolin-6 stimulated the food intake in aged mice. Contrariwise, the hormone that stimulates hunger (ghrelin) could not induce food intake temptations, thereby suggesting that endogenous neuropeptide downstream of the hunger hormone (ghrelin), however, the opioid receptors (δ) might be impaired in old age mice. Overall, it may be concluded that rubiscolin-6 encourages food intake in the old age mice with an intact ghrelin resistance (Miyazaki et al. 2014). Earlier, it has been demonstrated that opioid peptide rubiscolin-6 activity through oral administration was blocked on naltrindole, a δ-receptor antagonist. Therefore, it assumes that this peptide has a role in the intake of food through the activation of δ-opioid receptors. The orexigenic effect induced by this exorphin was further inhibited by COX-2 inhibitor (celecoxib). It was also found that the expression of COX-2 and lipocalin-type (L) PGD synthase was increased with this peptide. Therefore, it demonstrates that this peptide shows orexigenic activity in wild-type and Knockout mice (hematopoietic (H)-PGDS), however, no changes were observed in L-PGDS KO mice. The effect (orexigenic) of this exorphin was diminished by the deletion of DP (1) and Y (1) receptor gene essential for PGD (2) and neuropeptide, respectively. Therefore these findings lead to the conclusion that this exorphin is a potential candidate for encouraging the food intake through COX-2 and leptomeningeal L-PGDS, proceeded by Y (1) and DP (1) receptors, downstream of opioid (δ) receptor (Kaneko et al. 2012a, b). In a similar attempt, the same investigators revealed the role of the central opioid system in the maintenance of food intake. The activation of this receptors system besides regulating food intake suppresses the consumption of a high-fat diet. The authors revealed that this activity is controlled through melanocortin-corticotropin-releasing factor (MC-CRF) receptors, however, without any role of PGD (2) system (orexigenic). Rubiscolin after oral administration was found to decrease the intake of a high-fat diet and the effect was blocked on naltrindole administration. Therefore, these observations suggest that this RuBisCo-derived peptide decreased consumption of a high-fat diet possibly through activation of the central δ-opioid receptor system (Kaneko et al. 2014). Artemova and coworkers have suggested that food-derived exorphins have a potential role in the suppression of aggregation and encourage a recurrence of partially unfolded stress proteins. The commercially synthesized exorphins with sequences similar to gluten exorphin C and other opioid peptides like rubiscolin and hemorphin have shown protective effects in this perspective. These beneficial effects have been demonstrated through various spectroscopic methods including turbidity, fluorescence, dynamic light scattering, and circular dichroism. These researchers have found a dose-dependent effect (suppression) of light scattering intensity of aggregates of lactalbumin and alcohol dehydrogenase in the presence of these exorphins. The quantity of nanoparticles that show large radii (hydrodynamic) is shifted to lower ones, associated with prolonged time of aggregation. The exorphins in the refolding medium were depicted to help reactivation of carbonic anhydrase and increase the productivity of the carbonic anhydrase soluble protein. These findings, therefore, suggest that exorphins from food sources may be potential exogenous complements that may aid the endogenous defense mechanisms of humans under stress conditions (Artemova et al. 2010). The plant-derived exorphins have been demonstrated to have a role in cockroach escape reaction in increased temperature environments. The median effective dose (ED_{50}) was found to increase two times (stay time, insect) at elevated temperatures in addition to the duration and dynamics. Moreover, the ED_{50} was found to have an inverse relationship with the peptide length and a direct correlation with the affinity to the DOP. Among many plant-derived exorphins studied, the rubiscolin was found more active. Therefore, it leads to the conclusion that plant food-derived exorphins may have a potential role in the regulation (suppression) of the protective behavior in insects (Gritsaĭ et al. 2009).

The researchers have studied the delayed effect of rubiscolin-5 on the learning in rat pups. The peptide has been found to improve the progression of the trained foraging reflex. The generated data validated that this exorphin has a remarkable and prolonged effect on the adaptation (environmental) of young mammals (Dubynin et al. 2008). Researchers have found that oral (100 mg/kg) or intraperitoneal (10 mg/kg) administration of rubiscolin-6 shows anxiolytic effect in mice. The same was blocked through the administration of naltrindole (1 mg/kg), a δ-opioid receptor antagonist. Therefore, the manifestation of this effect (anxiolytic) is suggested to occur through the mediation of δ-opioid receptors. Moreover, other findings revealed the blockade of this effect by sigma (1) and dopamine D (1) antagonists, without any change occurring with dopamine D (2) antagonists. Therefore, these results establish that this rubiscolin-mediated anxiolytic effect occurs through sigma and dopamine (1) receptors downstream of the δ-opioid receptor (Hirata et al. 2007). Researchers have revealed the role of rubiscolin-6 (isolated from the enzymatic hydrolysate of RuBisCo) in the memory development in mice. These observations have revealed the effect after intracerebroventricular (3 nmol/mouse) and oral administration (100 mg/kg) administration of this peptide in mice. The amount of this exorphin that produces this effect (memory consolidation) is smaller compared to the optimal doses needed for the manifestation of analgesic activity. Moreover, this effect was blocked on pretreatment with naltrindole (delta antagonist) in mice, therefore linking activation of the delta-opioid receptor with rubiscolin-6 (Yang et al. 2003).

7.6.2 Soymorphins

It is well-established that opioids are linked with emotional behavior (Asakawa et al. 1998). Researchers have reported that intraperitoneal administration (3–10 mg/kg) of soymorphins (5, 6 and 7) elevated the time consumed in the open arms, therefore suggesting the anxiolytic activities of these peptides. Although, soymorphins-4 could not observe the anxiolytic effect, nevertheless, the soymorphins 5 and 6 showed a bell-shaped dose-response curve. When these three exorphins were administered orally, they enhanced the time spent in open arms, therefore, leads to the conclusion that these peptides are orally active exorphins with anxiolytic activities. Therefore it presents proof that these peptides are absorbed intact without being hydrolyzed in the gut. The dose required to show the anxiolytic effect through oral route is three times higher (10 mg/kg) compared to the intraperitoneal (3 mg/kg) administration. Moreover, for soymorphin-7 this dose was tenfold higher for the oral (30 mg/kg) compared to intraperitoneal (3 mg/kg) administration. Therefore, it is assumed that the large size of this peptide (7 amino acids) finds it difficult to transport across the gut and reach the blood and cross the blood–brain barrier (BBB). Furthermore, these three exorphins could not alter the overall entry in the increased plus-maze test. They could not alter locomotor activity under experimental conditions. Therefore, they were unable to change the antinociceptive effect under these conditions. Overall, the authors suggest that a dose needed to induce anxiolysis is different from analgesia (Ohinata et al. 2007).

Soymorphins suppressed the consumption of food when they were administered to the fasted mice. Further, they decreased the intestinal transit and the dosage was quite low as compared to needed for induction of anorexigenic effects. Both the effects induced by these exorphins were blocked on the administration of either naloxone or naloxonazine that are the antagonists of MOP. Raclopride that is an antagonist of serotonin 5-HT (1A) receptor blocked the inhibitory activity in the intestinal transit produced by these exorphins. Saclofen, the antagonist GABA (B), or dopamine D (2) receptors blocked the inhibitory activity in the intestinal transit produced by soymorphins. Moreover, the investigators checked the order of activation of GABA (B), D (2), and 5-HT (1A) receptors through the application of their agonists and antagonists. The inhibitory activity of a 5-HT (1A) agonist, 8-Hydroxy-DPAT hydrobromide on the intestinal transit was blocked by saclofen or raclopride. The D (2) agonist, bromocriptine induced a decrease in intestinal transit that was inhibited by saclofen and WAY100135 could not produce such a blockade. Moreover, the agonist of GABA (B) and baclofen decreased the intestinal transit could not be blocked by either raclopride or WAY100135. These observations, therefore, advocated that 5-HT (1A) activation produces D (2) preceded by GABA (B) activations in intestinal motility. In conclusion, the researchers found that soymorphins on oral administration decrease the food intake and intestinal transit through M1OP coupled to GABA(B), 5-HT(1A), and D(2) systems (Kaneko et al. 2010).

7.6.3 Oryzatensin

Takahashi and coworker in 1994 have found that rice albumin-derived oryzatensin has a possible role in promoting phagocytic activity in polymorphonuclear leukocytes (PMNLs) in humans. Additionally, the authors have established an increase in the generation of superoxide ions through this exorphin from peripheral leukocytes (human) (Takahashi et al. 1994). More recently in 2017 Barati and coworkers have depicted the role of oryzatensin in cancer progression in vitro. Peripheral blood mononuclear cells (PBMCs) were assessed for their proliferation and secretion of cytokines. Moreover, the HEP-G2 cell lines were cultured with PBMCs (oryzatensin stimulated) and checked for the expression of Matrix Metalloproteinase 9 (MMP-9) and vascular endothelial growth factor (VEGF). The cells (PBMCs) from healthy volunteers ($n = 25$) were isolated by density gradient (Ficoll) centrifugation. Real time-PCR and ELISA were used for the expression analysis studies cytokine secretion, respectively. Additionally, MTT assay was exploited for the assessment of cell proliferation. The authors found that oryzatensin stimulated cell proliferation (PBMCs), decreased the IFN-γ, and increased the IL-4 levels from these isolated cells. Moreover, real-time PCR showed a significant increase in the expression of VEGF and MMP-9 from HEP-G2 cell lines. Therefore, the authors demonstrated a possible potential role of this peptide in immunomodulation through the sensitization of cancer cells (Barati et al. 2017). Reports have already established anti-amnesic and anti-analgesic effects of complement C3a. This peptide that is an opioid antagonist has been found to get inactivated after oral consumption. A peptide was synthesized that resembles the rice oryzatensin at the carboxyl-terminal and demonstrates selectivity for complement C3a receptor. It has been established that the five amino acids at the carboxyl-terminal are minimally essential for the determination of C3a activity. In the previous section, we have seen that this peptide shows MOP selectivity. It was found that both the synthetic peptide and oryzatensin demonstrated analgesic effect when given through intracerebroventricularly in mice. Moreover, naloxone blocked the effect induced by these peptides. Additionally, it was found that the substitution of a hydrophobic amino acid by tyrosine diminished the activity. Therefore this synthetic peptide demonstrated the strongest C3a activity. Furthermore, this synthetic peptide reversed the analgesic effect induced by morphine when given at a dose of 300 mg/kg through oral administration. Also, this peptide was found to improve amnesia induced by scopolamine and ischemia (Jinsmaa et al. 2001). A synthetic exorphin [Trp5]-oryzatensin (5–9) (WPLPR) composed of five amino acids (pentapeptide) with a sequence similar to the carboxyl-terminal of oryzatensin (GYPMYPLPR) was synthesized. This synthetic peptide is an antagonist for complement C3a receptor. Mice (fasting) were administered with 3–30 nmol/mouse or 30–300 mg/kg of this synthetic peptide through intracerebroventricular or intraperitoneal routes, respectively. The results indicated that this exorphin decreased the food intake in these mice models. Moreover, oral administration of the peptide (300 mg/kg) also showed the same effect. Additionally, it was observed that after intraperitoneal administration of this peptide (300 mg/kg), it decreased the gastric emptying. The anorexigenic effect of this peptide was blocked by cyclooxygenase (COX) inhibitor. Moreover, the prostaglandin (PGE) receptor EP4 antagonist also blocked the activity. The investigators, therefore, conclude that this peptide has a role in the suppression of food intake possibly through PGE2 generation and EP4 receptor activation (Ohinata et al. 2007).

References

Artemova NV, Bumagina ZM, Kasakov AS et al (2010) Opioid peptides derived from food proteins suppress aggregation and promote reactivation of partly unfolded stressed proteins. Peptides 31 (2):332–338

Asakawa A, Inui A, Momose K et al (1998) Endomorphins have orexigenic and anxiolytic activities in mice. Neuroreport 9:2265–2267

Barati M, Yousefi M, Ebrahimi-Mameghani M (2017) Oryzatensin-stimulated pbmcs increase cancer progression in-vitro. Iran J Allergy Asthm 16(2):120–126

Bhat ZF, Kumar S, Bhat HF (2015a) Bioactive peptides from egg: a review. Nutr Food Sci 45:190–212

Bhat ZF, Kumar S, Bhat HF (2015b) Bioactive peptides of animal origin: a review. J Food Sci Technol 52:5377–5392

Carrasco-Castilla J, Hernández-Álvarez AJ, Jiménez-Martínez C et al (2012) Use of proteomics and peptidomics methods in food bioactive peptide science and engineering. Food Eng Rev 4 (4):224–243

Cassell RJ, Mores KL, Zerfas BL et al (2019) Rubiscolins are naturally occurring g protein-biased delta opioid receptor peptides. Eur Neuropsychopharmacol 29(3):450–456

Chesnokova E, Saricheva N, Dubynin V et al (2014) Behavioral effect of soymorphin-5-amide in rats. Mosc Univ Biol Sci Bull 69(3):103–107

Chiang T, Sansuk K, Van Rijn RM (2016) Beta-arrestins 2 dependence of delta opioid receptor agonists is correlated with alcohol intake. Br J Pharmacol 173:323–343

Dubynin VA, Malinovskaya IV, Belyaeva YA (2008) Delayed effect of exorphins on learning of albino rat pups. Biol Bull 35:43–49

Fields K, Falla TJ, Rodan K et al (2009) Bioactive peptides: signaling the future. J Cosmet Dermatol 8(1):8–13

Fukudome S, Yoshikawa M (1992) Opioid peptides derived from wheat gluten: their isolation and characterization. FEBS Lett 296(1):107–111

Garg S, Nurgali K, Mishra V (2016) Food proteins as source of opioid peptides-A review. Curr Med Chem 23(9):893–910

Gritsaĭ OB, Dubynin VA, Bespalova ZD et al (2009) Effects of several exorphins and endorphins on the escape reaction of the cockroach Periplaneta Americana under elevated temperature conditions. Zh Evol Biokhim Fiziol 45(4):391–407

Hirata H, Sonoda S, Agui S et al (2007) Rubiscolin-6, a δ opioid peptide derive from spinach RuBisCo, has anxiolytic effect via activating σ1 and dopamine D1 receptors. Peptides 28 (10):1998–2003

Jinsmaa Y, Takenaka Y, Yoshikawa M (2001) Designing of an orally active complement C3a agonist peptide with anti-analgesic and anti-amnesic activity. Peptides 22(1):25–32

Kairupan TS, Cheng KC, Asakawa A (2019) Rubiscolin-6 activates opioid receptors to enhance glucose uptake in skeletal muscle. J Food Drug Anal 27(1):266–274

Kaneko K, Iwasaki M, Yoshikawa M et al (2010) Orally administered soymorphins, soy-derived opioid peptides, suppress feeding and intestinal transit via gut μ1-receptor coupled to 5-HT1A, D2, and GABAB systems. Am J Physiol Gastrointest Liver Physiol 299(3):799–805

Kaneko K, Lazarus M, Miyamoto C et al (2012a) Orally administered rubiscolin-6, a δ opioid peptide derived from RuBisCo, stimulates food intake via leptomeningeal lipocallin-type prostaglandin D synthase in mice. Mol Nutr Food Res 56(8):1315–1323

Kaneko K, Yoshikawa M, Ohinata K (2012b) Novel orexigenic pathway prostaglandin D2-NPY system--involvement in orally active orexigenic δ opioid peptide. Neuropeptides 46(6):353–357

Kaneko K, Mizushige T, Miyazaki Y et al (2014) δ-Opioid receptor activation stimulates normal diet intake but conversely suppresses high-fat diet intake in mice. Am J Physiol Regul Integr Comp Physiol 306(4):265–272

Kitts DD, Weiler K (2003) Bioactive proteins and peptides from food sources. Applications of bioprocesses used in isolation and recovery. Curr Pharm Des 9:1309–1323

Lee SY, Hur SJ (2017) Antihypertensive peptides from animal products, marine organisms, and plants. Food Chem 228:506–517

Mitsumoto Y, Sato R, Tagawa N et al (2019) Rubiscolin-6, a δ-opioid peptide from spinach RuBisCo, exerts antidepressant-like effect in restraint-stressed mice. J Nutr Sci Vitaminol 65 (2):202–204

Miyazaki Y, Kaneko K, Iguchi S (2014) Orally administered δ opioid agonist peptide rubiscolin-6 stimulates food intake in aged mice with ghrelin resistance. Mol Nutr Food Res 58 (10):2046–2052

Möller NP, Scholz-Ahrens KE, Roos N et al (2008) Bioactive peptides and proteins from foods: indication for health effects. Eur J Nutr 47(4):171–182

Ohinata K, Agui S, Yoshikawa M (2007) Soymorphins, novel μ opioid peptides derived from soy β-conglycinin β-subunit, have anxiolytic activities. Biosci Biotechnol Biochem 71 (10):2618–2621

Robins MT, Chiang T, Mores KL et al (2018) Critical role for Gi/o-protein activity in the dorsal striatum in the reduction of voluntary alcohol intake in C57Bl/6 mice. Front Psych 9:112

Sánchez D, Vázquez A (2017) Bioactive peptides: a review. Food Qual Saf 1(1):29–46

Takahashi M, Moriguchi S, Yoshikawa M et al (1994) Isolation and characterization of oryzatensin: a novel bioactive peptide with ileum-contracting and immunomodulating activities derived from rice albumin. Biochem Mol Biol Int 33(6):1151–1158

Yamada Y, Muraki A, Oie M et al (2012) Soymorphin-5, a soy-derived μ-opioid peptide, decreases glucose and triglyceride levels through activating adiponectin and PPARα systems in diabetic KKAy mice. Am J Physiol Endocrinol Metab 302(4):433–440

Yang S, Yunden J, Sonoda S et al (2001) Rubiscolin, a δ selective opioid peptide derived from plant RuBisCo. FEBS Lett 509(2):213–217

Yang S, Kawamura Y, Yoshikawa M (2003) Effect of rubiscolin, a δ opioid peptide derived from RuBisCo, on memory consolidation. Peptides 24(2):325–328

Abstract

The proteins are isolated from foods like wheat, milk, soybean, spinach, and rice. The isolated proteins are dialyzed against weak salt solutions to remove the salts and sugars. The proteins are estimated by Folin Lowry or Bradford method. The purity of these proteins is checked by SDS-PAGE. Simulated gastrointestinal digestion (SGID) of the isolated proteins is done through proteolytic enzymes at 37 °C for different incubation time. The hydrolysates are filtered through centrifugal filters with different molecular weight cut-off filters (MWCOF). The generated exorphins are analyzed through analytical HPLC (RP-HPLC) using a C18 column and correlated with chemically synthesized standard exorphins. The peptide fractions are collected on the preparative HPLC and freeze lyophilized. The lyophilized powder is reconstituted in a buffer and the opioid activity is assessed on isolated organ bath using GPI or MVD assay. The other reconstituted fraction is estimated for the presence of exorphins by ELISA and the data is validated through mass spectrometers like a tandem, quadrupole ion-trap (QIT), or time of flight (TOF) (MALDI-TOF/TOF).

8.1 Introduction

In this chapter, a description of the analysis of food-derived exorphins is presented. The extraction of food proteins like casein, gluten, albumin, RuBisCo, and β-conglycinin is done from their corresponding foods. The extracted proteins are then dialyzed against dilute salt solutions to remove the unwanted salts used in extraction protocols. Proteins are quantified in the extracts with sensitive protein estimation methods. Sodium dodecyl sulfate-polyacrylamide gel electrophoresis (SDS-PAGE) is performed for purity analysis of proteins. Simulated gastrointestinal digestion (SGID) is performed with pepsin, trypsin, chymotrypsin, elastase, and leucine aminopeptidase (LAP). Hydrolysates are filtered through centrifugal filters

M. R. Ul Haq, *Opioid Food Peptides*, https://doi.org/10.1007/978-981-15-6102-3_8

with appropriate pore size. The filtrates with peptides of interest are analyzed on analytical HPLC corresponding to the standard chemically synthesized peptides. The fractions are collected corresponding to the retention time of standard peptides. For fraction collection, the hydrolysates are run on the preparative HPLC. The collected fractions are freeze lyophilized and the powder form is stored for further analysis. Some of the powder fraction is reconstituted in the buffer and checked for the opioid activity. The remaining fraction is subjected to amino acid sequencing on the sequencer. The concentration of the peptides is determined using ELISA or peak area on analytical HPLC (Raies et al. 2015).

8.2   Isolation of Proteins from Food Source

8.2.1   Isolation of Alpha, Beta, and Kappa Caseins from Cow Milk

Milk is collected from the cows, buffalos, or goats depending upon the type of analysis. The milk samples are centrifuged ($5000 \times g$ for 20 min at 4 °C) to remove the fat and the resultant is the skim milk. Casein is precipitated from this milk by isoelectric precipitation (Davies and Law 1977) at 37 °C. The pH is changed to 4.6 with HCl (1.0N) by continuous stirring using a glass rod. The pH adjusted is the isoelectric point of casein; i.e., at this pH, casein has the least solubility and therefore precipitates. The white casein precipitate so obtained is kept at room temperature for half an hour. The casein precipitate is packed in three to four layers of muslin cloth and kept under tap water for washing to remove the acid and other salts. The final washing is performed with warm water. The casein is taken and reconstituted with water equal to the initial volume of skim milk and pH adjusted with 1.0N NaOH to 6.9 with the help of magnetic stirrer. The re-precipitation of the casein is done with 1N HCl by decreasing pH to 4.6. The precipitate is again wrapped in three to four layers of muslin cloth and washed under tap water. Finally, the casein (wet) is dried with the use of acetone followed by ether. The dried casein is used fresh or stored at −20 °C until further analysis. The isoelectric casein is fractionated to alpha, beta, and kappa caseins (Fox and Guiney 1972; Zittle and Custer 1963). Beta-casein is used for the study of β-casomorphins and alpha-caseins for α-casomorphins. The whey is taken for analysis of lactalbumin and lactoglobulin.

For fractionation of various caseins from isoelectric casein, dissolve isoelectric casein in urea to 6.6 M and adjust pH to 1.5 with 7N H_2SO_4 and further dilute to get 2.2M urea and allow standing for 2 h and then centrifuged at 5000 g for 15 min. The supernatant contains the kappa casein and the sediment with alpha- and beta-casein. Take supernatant and perform SGID for analysis of casoxins. Take the sediments for the analysis of alpha- and beta-caseins. Take sediment and dissolve in urea to 6.6M and adjust pH to 4.5 and dilute it to 3.3 and allow standing for half an hour. Centrifuge the latter solution at $5000 \times g$ for 15 min. The sediment contains alpha-casein and is a source of alpha-casomorphins. Take supernatant and dilute to 1M urea and adjust pH to 4.9 followed by centrifugation at $5000 \times g$ for at least 15 min. Discard the supernatant and take sediment. The sediment contains crude beta-casein. To purify the same, follow resuspension of the sediment in 6.6M urea and adjustment of pH to 4.5. Dilute the same to 3.3M followed by centrifugation at 5000 × g for at least 15 min. Dilute the supernatant to 1M followed by adjustment to pH 4.9. Centrifuge again at 5000 × g for at least 15 min and collect the sediment, and finally, disperse in water and adjust pH to 7 followed by dialysis. Take beta-casein and perform SGID for the release of beta-casomorphins (Raies et al. 2015).

8.2.2   Milk Alpha-lactalbumin and Beta-lactoglobulin

Fractionate the milk into isoelectric casein and whey as discussed in the above section. Take the whey and perform salt precipitation with ammonium sulfate (50%), followed by dialysis through 1 kDa cut-off filters, and proceed for freeze-drying in a freezer. For this purpose, add finely powdered ammonium sulfate in the whey sample to provide 10–90% final saturation of the applied salt. Allow standing the samples overnight at 4 °C for the proteins to precipitate. Next day centrifuge the sample at 16,000 × g for 15 min at 4 °C. Dialyze the samples with dialysis tubing (3 kDa cut-off filters) and then freeze-dry the samples. Perform SDS-PAGE for purity analysis and affinity chromatography and HPLC for separating alpha-albumin and beta-lactoglobulin. The separated alpha-lactalbumin and beta-globulin from the whey are analyzed by SDS-PAGE. The separating gel (12%) and stacking gel (5%) are used for such analysis. The proteins are allowed to move through stacking gel at 80 V and in the separating gel at 120 V. The standard proteins are molecular ladder with different molecular weights depending on the analyte and the samples are run in parallel to the reference marker (Hattori et al. 2004). This will be followed by ion-exchange chromatography and RP-HPLC for purification (Cheison et al. 2011; Kong et al. 2012).

8.2.3   Milk Lactoferrin

Lactoferrin is isolated from the milk because it is a source of lactoferroxins as discussed in Chap. 4. Milk is collected from animals like cows, goats, sheep, and camels. Isolation of this iron-containing protein is done with cation exchange chromatography. Briefly, milk is centrifuged at 2500 × g at 4 °C for 30 min to remove fat and get the skim milk. Dilute the latter with equal volume of dilution buffer (NaH_2PO_4, 0.04M; NaCl, 0.8M; Tween 20, (v/v); 0.04% pH 7.4). Incubate the latter with SP-Sepharose (sulfopropyl strong cation exchanger) at 4 °C overnight. Next day, wash the SP-sepharose with phosphate buffer saline to remove the unbound proteins. Pack the gel in the column and elute lactoferrin with elution buffer at a flow rate of 3 ml/min (Van Berkel et al. 1995; Conesa et al. 2008). Perform dialysis for lactoferrin against ammonium bicarbonate (25 mM), lyophilized through freeze-drying, and store at −20 °C until use. Estimate the protein concentration with the Folin Lowry method or Bradford's Assay. Check the purity of the isolated lactoferrin with SDS-PAGE.

The alternate method is to take the whey (obtained through isoelectric precipitation of casein) and add sodium hydroxide (1N) with constant stirring until pH reaches 6.0. Proceed by addition of an equal volume of ammonium sulfate (45%) with slow stirring initially followed by little vigorous stirring and incubation at room temperature for 1 h. After completion of the incubation period, add hydrochloric acid (1N) slowly with constant stirring until pH reaches 4.0 and then increase the pH to 8.0 with sodium hydroxide. Subsequently, add an equal volume of ammonium sulfate (80%), initially stir slowly (100 rpm) and then increase (420 rpm), consume all the ammonium sulfate solution during stirring for about 1 h. Allow the sample to stand overnight at 4 °C. The next day, centrifuge the solution to get the precipitated lactoferrin. Dissolve the lactoferrin in phosphate buffer saline and dialyze against the low strength salt solution. Check the concentration of proteins by the Folin Lowry or Bradford method. Check the purity of the isolated lactoferrin with SDS-PAGE and then freeze lyophilize the proteins on the rotatory concentrator.

8.2.4   Wheat Glutenin and Gliadin

Take 100 mg wheat flour and add butanol (1 ml) and mix for about 1 h at 20 °C. Centrifuge the mixture and discard the supernatant. Take the residue and mix it with sodium chloride (1 ml, 0.5M) for about 1 h at 20 °C. Centrifuge and discard the supernatant. Take residue and wash with water and then add ethanol (70%, 1 ml) to extract gliadin. Take remaining residue and add propane (1 ml, 50%), 2-ME (2%) and acetic acid (1%). Incubate the latter for 1 h at 20 °C. Centrifuge the mixture after completion of the incubation period. Take supernatant that contains glutenin (Tatham et al. 2000). Check the concentration of proteins by the Folin Lowry or Bradford method. Check the purity of the isolated β-conglycinin with SDS-PAGE and then freeze lyophilize the proteins on the rotatory concentrator.

8.2.5   Spinach RuBisCo

Take spinach leaves (approximately 100 g) and grind it in presence of liquid nitrogen (frozen) to a fine powder in a mortar and pestle. Add extraction buffer (200 ml) and continue stirring for about half an hour while keeping accurate regards of temperature (keep below 0 °C). The next steps are performed at 4 °C. The extract is taken and filtered through Miracloth (two to three layers) and then centrifuge the filtrate at 48,000 × g for 15 min. Proceed by addition of ammonium sulfate (saturated, pH 7.0) in increments with continuous stirring to get a saturation of 35% and continue for an additional half an hour. Centrifuge the solution at 10,000 × g for 5 min and take the precipitate and resuspend in buffer A (5 ml) (10 mm BTP, 10 mm dithiothreitol, 0.2 mM ATP, 100 mM potassium chloride, pH 7.0) with ammonium sulfate (35%). Perform washing through centrifugation at 10,000 × g for 5 min. Take the pellet and resuspend in buffer A (3 ml) and then centrifuge. Take supernatant and remove salt on the Sephadex G25 column equilibrated with buffer A. The peak corresponding to the protein is collected and later freeze in liquid nitrogen. Thaw the extract and dilute with an equal volume of buffer (10 mM BTP, 0.2 mm ATP, and 10 mm dithiothreitol, pH 7.0). Follow centrifugation at 30,000 rpm for about 1 h. Take the supernatant and apply to the FPLC with about 10–15 mg/injection. Perform washing with buffer (4 mM dithiothreitol and 20 mM BTP, pH 7.0). Follow gradient elution from 0 to 500 mM potassium chloride and flow rate of 0.5 ml/min. Take the fractions and check for RuBisCo activity (Robinson et al. 1988). Dialyze the solution to remove the salts. Check the concentration of proteins by the Folin Lowry or Bradford method. Check the purity of the isolated RuBisCo with SDS-PAGE.

8.2.6   Soy Beta-conglycinin

The proteins in soy are first extracted by Panthee method. The sample is weighed and 10 g are ground in cold water (20 °C) till flour is of uniform size. The soluble protein fraction is extracted at room temperature for at least 1 h. The flour (1 g) is mixed with Tris-HCl buffer (0.2M, pH 8.0). The ratio of flour to buffer is 1:15 (w/v) supplemented with β-mercaptoethanol (0.1M). Centrifuge the mixture at $10 \times$ g at 4 °C for 10 min. Remove the fat layer and take 1 ml of supernatant. Add an equal volume of both SDS (5%) and β-mercaptoethanol (0.1M) to dissociate proteins in the extract and then warm it at 44.5 °C for 10 min in a water bath (Panthee et al. 2004; Delwiche et al. 2007).

Alternately, in Deak's method, 100 g of defatted flour (soy) is extracted with water (deionized) (1:15 w/v) and adjusting pH to 8.5 with sodium hydroxide (2N). The slurry is allowed to stand for 1 h with intermittent stirring. Centrifuge at 14,000 $\times$ g for half an hour at 15 °C. Take the supernatant (extract) and add sodium bisulfate and calcium chloride to get either calcium or sulfate ions to 5 mM. Adjust pH to 6.4 with hydrochloric acid (2N). Take the slurry and allow it to store for 12–16 h at 4 °C. Centrifuge the slurry at 14,000 $\times$ g 4 °C for 30 min. Take the precipitate (glycinin-rich) and neutralize. Take supernatant and adjust pH of 4.8 with hydrochloric acid. Allow it to stand for 1 h with intermittent stirring. Centrifuge at 14,000 $\times$ g for half an hour at 4 °C. Take the precipitate that is rich in β-conglycinin and repeat as above. Take supernatant (whey) and start freeze-drying till extract is thick. Store the samples in sealed containers at 4 °C until use. The yield total proteins and β-conglycinin by this method are 80 and 85.6%, respectively. Moreover, the β-conglycinin subunits are comprised of alpha subunit (38.0%) and α' subunit (27.3%) (Deak et al. 2007). β-Conglycinin protein in soybean is comprised of 6 isomers designated from B1 to B6. These can be dissociated through column chromatography with diethyl aminoethyl (DEAE) and carboxymethyl (CM)-Sephadex in buffers containing 6M urea. Estimate the proteins by Folin Lowry or Bradford method. Check the purity of the isolated β-conglycinin with SDS-PAGE and then freeze lyophilize the β-conglycinin on the rotatory concentrator.

8.2.7 Rice Albumin

Albumin is the source of oryzatensin and is isolated from the rice. Take rice bran (full-fat) and extract it with about tenfold volume of organic solvent like hexane for at least 4 h. Centrifuge the same for 10 min at 4000 rpm. Take supernatant and remove the hexane through evaporation to get the defatted rice bran. Take 100 g of defatted rice bran mixed with a half-liter of different solvents for at least 1 h on magnetic stirring. Follow the order of extraction as sodium chloride (0.4M), alcohol (80%), and sodium hydroxide (0.01M). Keep the extract separately and re-extract the residue. Follow centrifugation at 4500 rpm for 15 min at each extraction step. The subsequent extraction step is repeated to a residue with a half-liter of each extracting buffer. The individual extractions were pooled and then filtered. The filtrate with 0.4M sodium chloride is dialyzed against water for 3 days at 4 °C. Perform centrifugation at 4500 rpm for 15 min. Take supernatant with the protein of interest, i.e., albumin (Wang et al. 2014). Check the concentration of proteins by the Folin Lowry or Bradford method. Check the purity of the isolated RuBisCo with SDS-PAGE and then freeze lyophilize the albumin on the rotatory concentrator.

8.3 Dialysis of Extracted Proteins

Dialysis is a technique that allows the removal of salts, chemicals, and other unwanted molecules to diffuse out of the sample through a semipermeable membrane by the process of passive diffusion. In this procedure, two solutions are needed, sample and the buffer solution, the latter is usually 200–500 times larger in volume. The molecules with molecular size greater than the pore size of the semipermeable membrane are retained, the remaining are diffused out in the buffer solution. The buffer solution is changed three to four times when equilibrium starts to develop. In this manner, the unwanted chemicals or molecules are diffused out of the sample to the required levels. The filters called cut-off filters define the pore size of the membrane. It is measured in angstrom (Å) units. A larger cut-off has more pore size and vice versa. The molecular cut-off filter indicates the molecule with the smallest size that is unable to cross the membrane. For instance, a cut-off filter with 10 K MWCO will retain molecules with a molecular mass equal to or greater than 10 kDa. The filters are fitted with membranes that are commonly manufactured from regenerated cellulose. For dialysis of the extracted proteins mentioned in the above section, the carbohydrates and different salts needed for extraction are removed through dialysis using different molecular weight cut-off filters depending on the molecular mass of the extracted protein. Dialysis is usually performed at 4 °C for 24 h, with intermittent changing of the buffer.

8.4   Protein Estimation by Folin Lowry Method

The concentration of extracted proteins like caseins, alpha-lactalbumin, beta-lacto-globulin, glutenin, gliadin, beta-conglycinin, spinach RuBisCo, and rice albumin from various food sources is checked by Lowry et al. (1951).

Reagents
- Solution A: Copper sulfate solution (0.5%)
- Solution B: Sodium potassium tartrate (1.0%)
- Solution C: Sodium carbonate (2.0%) in sodium hydroxide (0.1N)
- Lowry's Reagent: This reagent is prepared by mixing A, B, and C solutions in the ratio of 1:1:48
- Bovine Serum Albumin (BSA) (1 mg/ml)
- Folin Ciocalteu's phenol reagent (1N): Make fresh and dilute 1:1 with distilled water

Procedure
Different aliquots from 20 to 100 µl are added in the tubes from S1 to S5. These are the standards with known concentration. A tube without a sample or standard protein serves as the blank. The BSA solution (1 mg/ml) serves as the standard. The samples (20 µl) of various extracted proteins are added in test tubes. All the test tubes including the blank, standards, and samples are levelled to 1 ml with distilled water. Lowry's reagent (5 ml) is added to each test tube including blanks, standards, and samples. The tubes are then mixed and incubated for 10–15 min at room temperature. After completion of incubation, add 0.5 ml 1N Folin and Ciocalteu's phenol reagent. The solution is again mixed well and incubated at room temperature for 30 min in dark. Read absorbance at 660 nm. Plot a standard curve using the known BSA concentration (S1–S5). Determine the unknown concentration of extracted proteins (samples) from the standard curve.

8.5   Purity Analysis by SDS-PAGE

The purity of the extracted proteins like caseins, alpha-lactalbumin, beta-lactoglob-ulin, glutenin, gliadin, beta-conglycinin, spinach RuBisCo, and rice albumin from various food sources is checked by SDS-PAGE (Laemmli 1970).

Reagents
- Stock acrylamide/bisacrylamide (30%; 2.66% C): Weigh about 29.2 g of acrylamide and 0.8 g of bisacrylamide and dissolve in water. Make the final volume to 100 ml.
- Tris-HCl: Prepare 1.5M tris-HCl and adjust pH to 8.8
- Tris-HCl: Prepare 0.5M tris-HCl and adjust pH to 6.8
- Sodium dodecyl sulfate (SDS): Prepare 10% SDS solution in distilled water
- Ammonium persulfate solution (APS): Prepare 10% APS in distilled water

Table 8.1 Preparation of gel solutions (Raies et al. 2015)

Reagents	Unit	Stacking gel 4%	Separating gel 15%
30% Acrylamide/bis acrylamide	ml	1.3	7.5
Tris-HCl (0.5M, pH 6.8)	ml	2.5	–
Tris-HCl (1.5M, pH 8.8)	ml	–	3.75
SDS (10%)	ml	0.1	0.15
APS (10%)	ml	0.1	0.15
TEMED	ml	0.01	0.006
Distilled water	ml	5.6	3.45

- Electrode buffer ($5\times$, pH 8.3): Weigh 7.5 g of tris base, 2.5 g of SDS, and 36 g of glycine. Dissolve these three components in a little volume of water (300 ml) and make the final volume to 500 ml with distilled water. Adjust the pH of this buffer to 8.3. This buffer is $5\times$ and stored at 4 °C until further use. Just before the use, dilute the buffer five times to get a working buffer with $1\times$ buffer strength. This working buffer has tris (25 mM), SDS (0.1%), and glycine (192 mM).
- Sample buffer ($2\times$): Sample buffer (10 ml) is made by mixing bromophenol blue (2.5 mg), 1.6 ml glycerol (20%), 3.2 ml of 10% SDS (4%), 0.8 ml beta-mercaptoethanol (10%), and 1 ml of 0.5M Tris-HCl (pH 6.8). The total volume is made 10 ml with distilled water and stored at 4 °C till further use.
- Staining solution: Weigh 100 mg of CBB-R250 and dissolve in 10 ml of water. And add to it methanol (50 ml) and glacial acetic acid (40 ml). Filter this solution and then store it at room temperature.
- Destaining solution: This solution is made as 10% of acetic acid.
- Gel solutions: The stacking and separating gels are prepared as shown in Table 8.1.

Gel Preparation

The gel plates (8 × 7 cm) are washed and assembled on the SDS casting system putting in between 1 mm spacers. Pour separating or running gel between the plates of about 3–4 cm and immediately add water at the top with the help of a syringe until this gel polymerizes. After polymerization, decant the water and add a stacking gel of about 2–3 cm and simultaneously place the comb in the stacking gel and allow it to polymerize. After polymerization, remove the comb and place the prepared gel plates into the gel unit and pour the electrode buffer to both lower and upper chambers of the SDS-PAGE assembly.

Sample Preparation

The protein samples like caseins, alpha-lactalbumin, beta-lactoglobulin, glutenin, gliadin, beta-conglycinin, spinach RuBisCo, and rice albumin should be cleared off insoluble materials. Mix 1:1 volume of sample and the sample buffer to the final concentration 1 µg/µl. Heat the solution on boiling water for 10 min followed by cooling immediately.

Electrophoresis
Take approximately 15–20 µg of the protein samples and load them into the wells. Perform electrophoresis initially at 40 V until sample crosses stacking gel, then increase the voltage to 80 V in the separating gel. Keep accurate regard for the temperature.

Gel staining and Destaining
After the completion of electrophoresis detected through tracking dye when it has moved about 80–90% of gel, remove the gel and proceed for staining. Put the gel in the staining solution for about 45–60 min. Remove the gel and proceed for destaining till the background is clear and the protein bands appear clear. The aforesaid processes (staining and destaining) are carried out on a shaker.

8.6   Simulated Gastrointestinal Digestion

The isolated proteins including caseins, alpha-lactalbumin, beta-lactoglobulin, glutenin, gliadin, beta-conglycinin, spinach RuBisCo, and rice albumin from various food sources is hydrolyzed through various enzymes simulating the gastrointestinal digestion (Raies et al. 2015).

Reagents
Hydrochloric acid (1N)
   Sodium Hydroxide (1N)

Enzymes
Pepsin, Trypsin, Chymotrypsin, Elastase, Leucine aminopeptidase, Pancreatin

Procedure
Dissolve isolated proteins from food sources in distilled water and adjust the concentration to 24 mg/ml. Adjust the pH to 2.5 with hydrochloric acid (1N). Add pepsin to this protein solution in the ratio of $E/S = 1/100$, w/w. Incubate the protein: enzyme solution for 5 h at 37 °C. The reaction is stopped by increasing the pH to 7.0 with sodium hydroxide (1N). Divide this pepsin hydrolysate into two fractions. Designate one as pepsin hydrolysate (P). Take another fraction of this pepsin hydrolysate and add to it trypsin in the ratio of enzyme: trypsin ($E/S = 1/100$, w/w) and designate it as (PT). Carry the reaction for 5 h at 37 °C. Stop the reaction by boiling. Divide this hydrolysate (PT) into two fractions and one fraction designated as PT. Take the second fraction and add to it chymotrypsin for 5 h at 37 °C in the ratio of ($E/S = 1/100$, w/w). Stop the reaction by boiling and designate the fraction as (PTC). Divide the latter into two fractions. Take one fraction and add porcine pancreas in the ratio of ($E/S = 1:100$) for 1 h at 37 °C and denote it as (PTEC). Stop the reaction by boiling. Divide the reaction into two fractions and to one fraction add leucine aminopeptidase in the ratio of ($E/S = 1:100$) for 16 h at 37 °C and denote it as (PTCEL) (Raies et al. 2015).

8.7   Ultrafiltration of Hydrolysates

The protein hydrolysates (P, PT, PTC, PTCE, and PTCEL) are filtered through different molecular weight filters depending on the pore size of the filters. The filtration is performed through centrifugation at 7500 × g for 10–20 min. Take filtrates and store them at 4 °C if used shortly, and store at −20 °C when analysis to be formed after a long time.

8.8   Analysis of Opioid Peptides

8.8.1   Detection on Analytical HPLC

The protein hydrolysates (discussed in the above section) after ultrafiltration are analyzed on the analytical reverse phase-high pressure liquid chromatography (RP-HPLC). The column is usually C18 sepharose for hydrophobic peptides. The reagents should be HPLC grade. The water (Millipore) should be filtered (0.22 μm filters) using filtration assembly and degassed to remove the dissolved gases.

Solvents
Solvent A: trifluoroacetic acid (0.1%) in degassed water (Millipore)
Solvent B: trifluoroacetic acid (0.1%) in water (Millipore), acetonitrile (90%), HPLC grade water (9.9%).

Operating Conditions (Raies et al. 2015)
System: Waters [32] HPLC system
   Pumps: 515 HPLC pump
   Software: Millenium[32] software
   Detector: 2487 dual λ absorbance detector
   Column: C_{18} Spherisorb®, ODS 4.6 × 250 mm, 5 μm particle size, 300 Å pore size
   Guard column: ODS2, 10 mm × 1.6 mm
   Flow rate used: 1 ml/min
   Volume injected: 20 μl
   Wavelength: 220 and 280 nm

HPLC Program
The program followed for analytical RP-HPLC is given in Table 8.2

Procedure
The HPLC program is started with the washing of the column with solvent A until a baseline is flat, i.e., there are no peaks on the chromatogram. This is followed by loading the standard opioid peptides usually 20 μl (chemically synthesized) via the injector loop to the column with the help of a syringe. Record the retention time (minutes) of the standard peptides and proceed for the application of the protein

Table 8.2 Analytical HPLC program

Time (min)	Solvent A (%)	Solvent B (%)
0.0	100.0	00.00
5.0	100.0	00.00
35.0	50.00	50.00
65.0	0.00	100.00
75.0	100.0	00.00

hydrolysates on the same analytical HPLC. Check for the presence of the peaks corresponding to the retention time of the standard peptides. Perform the detection on dual-wavelength including 220 and 280 nm.

8.8.2   Peptide Fraction Collection on Preparative HPLC

After detection of peaks in the protein hydrolysates corresponding to the standard peptides on analytical HPLC and then taken to preparative HPLC and run again. This is done due to the reason that the volume of the sample used in the latter is large and in turn, a sufficient volume of peptide fractions is collected for further analysis.

Solvents
Solvent A and Solvent B: The preparation and composition of these solvents are usually the same as described above for analytical HPLC.

Operating Conditions
System: AKTA purifier
  Pumps: 515 HPLC pump
  Software: Millenium^{32} software
  Detector: 2487 dual λ absorbance detector
  Column: RESOURCE™ RPC, particle size 15 μm
  Guard column: ODS2, 10 mm × 1.6 mm
  Flow rate used: 1 ml/min
  Volume injected: 1 ml
  Wavelength: 214 nm

HPLC Program
The HPLC program that is used for preparative HPLC is the same as given above (Table 8.2)

Procedure
The HPLC procedure is the same as discussed for analytical HPLC. The detection is done at 220, 280, or 214 nm. The fractions will be collected at the retention time corresponding to the standard peptides.

8.8.3   Freeze Lyophilization

The fractions collected on the preparative RP-HPLC are then frozen at $-20\,°C$. After freezing taken to the Centrifugal Vacuum Concentrator to remove the solvents and form powder at a pressure usually 76 mmHg and temperature (trap) of $-85\,°C$. The powder form of the peptides is either studied directly for further measurements or stored at $-20\,°C$ until further use.

8.8.4   Opioid Activity Determination

The opioid activity of food-derived exorphins may be checked through various pharmacological techniques. The commonly employed assays include the naloxone reversible inhibition of enzyme activity (adenylate cyclase) and electrically stimulated contraction of tissue preparations of guinea pig ileum or mouse vas deferens (Sharma et al. 1975; Hughes et al. 1975). In addition to these assays, the radioreceptor assay (receptor binding assay) is also exploited for this purpose (Pert and Snyder 1973). The opioid receptors are categorized in the G protein-coupled receptors family (GPCRs). Whenever there is an activation of these receptors it causes activation and inactivation of K^+ and Ca^{2+} channels, respectively. Moreover, their activation also leads to the inhibition of adenylate cyclase responsible for the formation of cyclic AMP from ATP (Satoh and Minami 1995). Many investigators have exploited the inhibition assay of adenylate cyclase through GPCRs for the determination of the opioid activity of exorphins released from casein and gluten (Zioudrou et al. 1979; Loukas et al. 1983). Whenever an exorphin binds to its corresponding opioid receptor in either brain membranes or hybrid cells, it results in the inhibition of adenylate cyclase that in turn decreases the concentration of cAMP assessed through various assays (Sharma et al. 1975; Satoh and Minami 1995). Moreover, for the determination of the opioid activity of these exorphins, the tissues like guinea pig ileum or mouse vas deferens are first stimulated electrically, then opioid peptide binds to their corresponding receptors on this tissue and leads to the inhibition of the contractions (Janecka et al. 2004; Teschemacher 2003). The food exorphins bind selectively with MOP and DOP on the guinea pig ileum and mouse vas deferens preparations, respectively. In this way, these assays are employed from the assessment of the activity of various exorphins (Janecka et al. 2004; Zioudrou et al. 1979). The receptor binding assay includes competition and saturation kinetics, and in the latter, the affinity of various food-derived exorphins for their corresponding receptors is evaluated. The former assay is used independently or subsequently to validate or confirm the results (Janecka et al. 2004). The disadvantage with the binding assay is the use of radioisotopes that are hazardous to use, need permissions, are expensive, and further, their disposal is an issue. Moreover, the food-derived peptides that will alter the binding of a selected molecule allosterically may not essentially enhance the signaling potency. Furthermore, the generated data from various laboratories are difficult to compare due to changes in synaptosomal preparations, the doses, radioligand employed, and the assay used for reporting this binding. In conclusion, the bioassays that include tissue preparations and adenylate cyclase are more advantageous compared to receptor binding methods, this is due to the reason that these are more feasible, less expensive, and demonstrate the very low effect on the environment. Fukudome and Yoshikawa in 1992 isolated four opioid exorphins from the wheat gluten. The peptides were named as gluten exorphin A5 (Gly-Tyr-Tyr-Pro-Thr), gluten exorphin A4 (Gly-Tyr-Tyr-Pro), gluten exorphin B5 (Tyr-Gly-Gly-Trp-Leu), and gluten exorphin B4 (Tyr-Gly-Gly-Trp). The opioid activity of four isolated wheat-derived exorphins (A/B) was checked through GPI and MVD assays. In MVD assay, the mice were sacrificed and the vas deferens collected, the tension was applied to the isolated tissue (vas deferens) through suspension of 0.02 g weight in a Magnus tube that was supplemented with Mg^{2+} free Krebs-Ringer solution and stimulated. In the case of GPI assay, the ileum is taken and tension applied with 0.5 g in Krebs-Ringer solution and stimulated. The contraction was documented with a transducer (isometric). These experiments of assessment of opioid activity were performed in the presence of L-leucyl-L-leucine (2 mM) together with bestatin (30 µM), thiorphan (0.3 µM), and captopril (10 µM). The IC_{50} values are the concentration of these opioid peptides in the hydrolysates that inhibits the stimulated (electrically) muscle contraction by 50%. The assessment of opioid activities of these peptides through the MVD assay was done in the presence and absence of 10^{-6} µM naloxone. The opioid activities that were reversed on the addition of naloxone and blocked by pretreatment with this compound were regarded as opioid activities. The radioreceptor assays were done as per the protocols of Pert and Snyder (1973) in presence of [^{3}H] DADLE (1 nM, 30.0 Ci/mmol) and [^{3}H] DAGO (1 nM, 47.8 Ci/mmol) for DOP and MOP and rat brain membrane. This radioisotope assay was performed for κ-receptors in presence of [^{3}H] EKC (1 nM, 24.5 Ci/mmol) and guinea pig cerebellum membrane. These opioid activity determination assays were done in the presence of inhibitors. The IC_{50} value is the concentration of the opioid peptide that will inhibit the binding of the labeled ligand by 50% (Fukudome and Yoshikawa 1992). Moreover, the researchers showed that the sequence of gluten exorphin A5 is located at 15 sites in the primary of glutenin that is high molecular weight protein found in wheat. This exorphin showed high selectivity for delta opioid receptor. Moreover, the structure-function relationship established that this exorphin is unique in that it possesses glycine at the amino-terminal unconventional to the presence of tyrosine in other food-derived opioid peptides. The wheat gluten derived B4 exorphin showed the presence of [Trp^{4}, Leus^{5}] enkephalin, which was found to be the most potent among all purified exorphins with IC_{50} values of 0.05 µM and 0.017 µM on GPI and MVD tissues, respectively (Fukudome and Yoshikawa 1992).

Reagents
- *Tyrode solution*: Sodium chloride (137 mM), potassium chloride (2.7 mM), calcium chloride (1.8 mM), magnesium chloride (0.1–1.0 mM), sodium bicarbonate (11.9 mM), sodium dihydrogen phosphate (0.4 mM), glucose (11.1 mM).

- *Modified Krebs-Ringer solution*: Sodium chloride (125 mM), potassium chloride (5 mM), potassium dihydrogen phosphate (1 mM), magnesium sulfate (1 mM), calcium chloride (1 mM), glucose (5.5 mM), and HEPES (20 mM).

Procedure

Animals, like mice, rats, etc., are usually kept in the Animal House of the university, pharmaceutical company, research institutes, etc. The animals are generally maintained in propylene cages and fed with proper diet and kept at room temperature under sterilized conditions. The methods and procedures applied for the measurement of opioid activity through animal tissue should be approved by the Institutional Animal Ethics Committee (IAEC). The opioid activity of the hydrolysates through various enzyme combinations (P, PT, PTC, PTCE, and PTCEL) is assessed on mouse vas deferens, rat ileum, or other tissues using isolated organ bath. The assembly consists of a water bath that is filled with water and temperature is maintained at 37 °C with the help of a thermostat. The bathtub is washed five to six times and then filled with the Tyrode solution, Krebs-Ringer solution, or any other solution before the tissue is mounted. The oxygen source is controlled through an oxygen regulator and the flow rate of oxygen is kept about 20–30 bubbles/min.

Dissection and Preparation of Tissue

The animals are kept fasting overnight and next day they are sacrificed. The tissue containing receptors for opioid peptides like ileum or vas deferens is removed and kept in a Petri dish filled with tyrode, Krebs-Ringer solution, or any other solution and washed with this solution. On the one end of the tissue, a thread is tied and kept close to the oxygen supply. The other end of the tissue is tied with a lever. The tissue including the oxygen tube is inserted in the bathtub with a filled tyrode or Krebs ringer solution.

Measurement of Opioid Activity

The tissues, like ileum, vas deferens, etc., are first allowed to relax for about 1 h. The alterations induced in the amplitude of contractions in the ileum or vas deferens are monitored and recorded through an isotonic transducer. Initially, a baseline is created through the spontaneous contractions and then samples are applied and followed by washing with the buffer solutions. Next, a mixture (sample and antagonist/agonist) is added to analyze the opioid activity. The alterations in the amplitude of the contractions of tissue are indicated as the responses. A dose-response curve is plotted between the concentration of opioid peptides and percentage response. A final value in terms of IC_{50} is interpreted for opioid activity.

8.8.5 Mass Spectroscopy of Peptide Fractions

The peptides in the form of powder (freeze lyophilized) that showed activity from the isolated organ bath or any other opioid activity determination method are validated for the presence of these peptides through sequencing. The samples are sequenced through mass spectrometers like a tandem, quadrupole ion-trap (QIT), or time of flight (TOF) (MALDI-TOF/TOF), for example, 4800 Proteomics Analyzer (AB Sciex). The spectra are generated to check peptides of interest using software, for example, Mascot sequence matching software (Matrix Science) with appropriate databases like Ludwig NR Database (Raies et al. 2015).

The sensitivity of the detection of the opioid and related peptides is enhanced by coupling the HPLC with tandem, quadrupole ion-trap (QIT), or time of flight (TOF) mass spectrometry (MS). The tandem MS has been already exploited for analysis of bioactive peptides like BCM-5 and BCM-7 from cheese (cheddar) manufactured through neutral protease obtained from Bacillus subtilis, infant milk, and casein variants (De Noni 2008; Haileselassie et al. 1999), BCM-9 (Italian goat cheese) (Rizzello et al. 2005), gluten exorphins A4, A5, B4, and B5 in cerebrospinal fluid (Fanciulli et al. 2006, 2007; Pennington et al. 2007). QIT-MS is even more sensitive, easy to use, and a time-saving technique used for the structural interpretation of opioids and related peptides. This mass spectrometer conveys information regarding the dominant productions generated to make fragmentation pathways and is already successful in the recognition of β-BCMs in milk and cheese (Juan-García et al. 2009). The advantage of using the TOF mass spectrometer is high-resolution power. Moreover, it is a more accurate method for the identification and validation of opioids and related peptides in a hydrolysate fraction. It acts by linking the elemental formula data from accuracy methods (high mass) and structural data obtained from fragmentation experiments. This mass spectrometer has already been exploited for the analysis of BCMs in Gouda cheese (Toelstede and Hofmann 2008) and other opioid peptides from bovine β-casein (Schmelzer et al. 2004, 2007).

8.8.6   In Silico Approach

This methodology is adopted for predicting the detection of opioid and other peptide sequences in the hydrolysates. The hydrolysis approaches were discussed in the previous sections, the protein is first hydrolyzed and then detected for the presence of opioid activity. The active fractions are then used for isolation and purification of opioids and other peptides in these hydrolysates. These methods are time consuming and wearisome. Additionally, there is a synergistic and additive effect of diverse elements (Udenigwe 2014). In this perspective, the bioinformatics plays an important role in this research area. It is time saving, direct, and precise procedures for the analysis of opioid and other food-derived peptides. In these methods, computer-enabled simulations have been exploited for determining the presence of food-derived exorphins or other peptides in various protein sources (Udenigwe et al. 2013; Carrasco-Castilla et al. 2012; Lacroix and Li-Chan 2012; Holton et al. 2013). Already some peptide databases like PEPBANK and BIOPEP are in place for peptide sequence elucidations from the protein database including Swiss-Prot, UniProtKB, and TrEMBL (Udenigwe et al. 2013). Moreover, other tools like proteolysis tools including ExPASy Peptide Cutter, BIOPEP, and PoPS are then exploited to analyze the specificities of numerous enzymes to produce the opioid and other peptides from various food sources. Already this method has been exploited for the prediction of ACE inhibitory peptides (Cheung et al. 2009; Dellafiora et al. 2015) and DPP-IV inhibitors (Lacroix and Li-Chan 2012; Nongonierma et al. 2014) from various foods and then validated through in vitro methods. The proteins from cereals more accurately demonstrate the presence of high frequencies of opioid, ACE inhibitory, DPPIV-inhibitory, anti-oxidant, anti-thrombotic, and hypotensive activities (Cavazos and Gonzalez de Mejia 2013).

8.8.7 Estimation of Opioid Peptides by ELISA

The opioid peptides or exorphins (alpha and beta-casomorphins, lactorphins, lactoferroxins, casoxins, gluten exorphins, gliadorphins, soymorphins, rubiscolins, and oryzatensin) released from the food proteins are estimated by ELISA. These kits differ in the sensitivity range. In our laboratory, we have checked the release of BCM-5 and BCM-7 with ELISA kits (Bachem Americas Inc., Torrance, CA) and the sensitivity range was 0.1–50 ng/ml (Raies et al. 2015). The working methodology should strictly be followed as given by the manufacturer. The principle could vary from the competitive to non-competitive immunoassay.

Reagents
- Enzyme-linked immunoassay buffer: The buffer that is provided with the kits has to be diluted in distilled water as per the manufacturer's instructions. The diluted solution is mixed well and brought to room temperature for 1 or 2 h.
- Standard solutions: The standard peptides (according to the exorphin to be estimated) that are provided with the ELISA kit should be reconstituted in ELISA buffer according to the manufacturer's instructions. The solutions are then allowed to the vortex and kept at room temperature before use.
- Antiserum: The ELISA kit is provided with the antiserum (lyophilized form) has to be reconstituted in 5 ml ELISA buffer. The solutions are then allowed to the vortex and kept at room temperature before use.
- Biotinylated-tracer: The tracer provided with the ELISA kit (lyophilized form) has to be reconstituted in 5 ml ELISA buffer. The solutions are then allowed to the vortex and kept at room temperature before use.
- Streptavidin-HRP: The vial supplied with the ELISA kit having SA-HRP is allowed to spin to collect the contents at the bottom of the vial. Dilute it according to the manufacturer's instructions and then vortex. Make this solution fresh like 10 min before use.

Procedure
- The ELISA plate supplied with the ELISA kit may be precoated or non-coated.
- For example, for the coated ELISA plates add 50 µl of the standard peptides in the standard wells supplied with the kit and reconstituted before the start of the experiment.

- In the sample wells add 50 µl of the reconstituted lyophilized extract fraction taken from different protein hydrolysates.
- In the blank wells add 50 µl EIA buffer.
- Add antiserum (25 µl) to each well.
- Incubate the plates at room temperature for 1 h.
- Add Bt-tracer (25 µl) to all the wells including sample, standard, and blank.
- Incubate for 2 h at room temperature.
- Wash plates five to six times with ELISA buffer.
- Add to each well SA-HRP including the sample, standard, and blank.
- Incubate the plate at room temperature for 1 h.
- Wash plate five to six times with ELISA buffer.
- Add to each well 100 µl substrate TMB (3,3′,5,5′-tetramethylbenzidine), including the standard, sample, and blank.
- Incubate for 15 min at room temperature.
- When color develops, stop the reaction after adding 2N HCl (100 µl) per well.
- Read absorbance at 450 nm.

References

Carrasco-Castilla J, Hernández-Álvarez AJ, Jiménez-Martínez C (2012) Use of proteomics and peptidomics methods in food bioactive peptide science and engineering. Food Eng Rev 4 (4):224–243

Cavazos A, Gonzalez de Mejia E (2013) Identification of bioactive peptides from cereal storage proteins and their potential role in prevention of chronic diseases. Compr Rev Food Sci 12 (4):364–380

Cheison SC, Lai MY, Leeb E et al (2011) Hydrolysis of β-lactoglobulin by trypsin under acidic pH and analysis of the hydrolysates with MALDI-TOF-MS/MS. Food Chem 125:1241–1248

Cheung IW, Nakayama S, Hsu MN et al (2009) Angiotensin-I converting enzyme inhibitory activity of hydrolysates from oat (Avena sativa) proteins by in silico and in vitro analyses. J Agric Food Chem 57(19):9234–9242

Conesa C, Sánchez L, Rota C et al (2008) Isolation of lactoferrin from milk of different species: calorimetric and antimicrobial studies. Comp Biochem Physiol 150:131–139

Davies DT, Law AJR (1977) An improved method for the quantitative fractionation of casein mixture using ion-exchange chromatography. J Dairy Res 44:213–221

De Noni I (2008) Release of beta-casomorphins 5 and 7 during simulated gastro-intestinal digestion of bovine beta-casein variants and milk-based infant formulas. Food Chem 110(4):897–903

Deak NA, Murphy PA, Johnson LA (2007) Characterization of fractionated soy proteins produced by a new simplified procedure. J Am Oil Chem Soc 84:137–149

Dellafiora L, Paolella S, Dall'Asta C et al (2015) Hybrid in silico/in vitro approach for the identification of angiotensin I converting enzyme inhibitory peptides from Parma dry-cured ham. J Agric Food Chem 63(28):6366–6375

Delwiche SR, Pordesimo LO, Panthee DR et al (2007) Assessing glycinin (11S) and β-Conglycinin (7S) fractions of soybean storage protein by near-infrared spectroscopy. J Am Oil Chem Soc 84:1107–1115

Fanciulli G, Azara E, Wood TD et al (2006) Quantification of gluten exorphin A5 in cerebrospinal fluid by liquid chromatography–mass spectrometry. J Chromatogr B 833(2):204–209

Fanciulli G, Azara E, Wood TD et al (2007) Liquid chromatography–mass spectrometry assay for quantification of Gluten Exorphin B5 in cerebrospinal fluid. J Chromatogr B 852(1):485–490

Fox PF, Guiney J (1972) A procedure for the partial fractionation of αS-casein complex. J Dairy Sci 39(1):49–53

Fukudome S, Yoshikawa M (1992) Opioid peptides derived from wheat gluten: their isolation and characterization. FEBS Lett 296:107–111

Haileselassie SS, Lee BH, Gibbs BF (1999) Purification and identification of potentially bioactive peptides from enzyme-modified cheese. J Dairy Sci 82(8):1612–1617

Hattori M, Miyakawa S, Ohama Y et al (2004) Reduced immunogenicity of β-lactoglobulin by conjugation with acidic oligosaccharides. J Agric Food Chem 52:4546–4553

Holton TA, Pollastri G, Shields DC et al (2013) CPPpred: prediction of cell penetrating peptides. Bioinformatics 29(23):3094–3096

Hughes J, Smith TW, Kosterlitz HW et al (1975) Identification of two related pentapeptides from the brain with potent opiate agonist activity. Nature 258(5536):577–580

Janecka A, Fichna J, Janecki T (2004) Opioid receptors and their ligands. Curr Top Med Chem 4:1–17

Juan-García A, Font G, Juan C et al (2009) Nanoelectrospray with ion-trap mass spectrometry for the determination of beta-casomorphins in derived milk products. Talanta 80(1):294–306

Kong XY, Wang J, Tang YJ et al (2012) HPLC analysis of α-lactalbumin and β-lactoglobulin in bovine milk with C4 and C18 column. J Northeast Agric Univ 19:76–82

Lacroix IM, Li-Chan EC (2012) Evaluation of the potential of dietary proteins as precursors of dipeptidyl peptidase (DPP)-IV inhibitors by an in silico approach. J Funct Foods 4(2):403–422

Laemmli VK (1970) Cleavage of structural proteins during the assembly of the head of bacteriophage T4. Nature 227(5259):680–685

Loukas S, Varoucha D, Zioudrou C et al (1983) Opioid activities and structures of alpha-casein-derived exorphins. Biochemistry 22:4567–4573

Lowry OH, Rosebrough NF, Farr AC, Randall I (1951) Protein measurement with Folin-phenol reagent. J Biol Chem 193(1):265–275

Nongonierma AB, Mooney C, Shields DC et al (2014) In silico approaches to predict the potential of milk protein-derived peptides as dipeptidyl peptidase IV (DPPIV) inhibitors. Peptides 57:43–51

Panthee DR, Kwanyuen P, Sams CE et al (2004) Quantitative trait loci for β-conglycinin (7S) and glycinin (11S) fractions of soybean storage protein. J Am Oil Chem Soc 81:1005–1012

Pennington CL, Dufresne CP, Fanciulli G et al (2007) Detection of gluten exorphin B4 and B5 in human blood by liquid chromatography-mass spectrometry/mass spectrometry. Open Spectrosc J 1:9–16

Pert CB, Snyder SH (1973) Opiate receptor: demonstration in nervous tissue. Science 179 (4077):1011–1014

Raies MH, Kapila R, Kapila S (2015) Release of β-casomorphin-7/5 during simulated gastrointestinal digestion of milk β-casein variants from Indian crossbred cattle (Karan fries). Food Chem 1 (168):70–79

Rizzello CG, Losito I, Gobbetti M et al (2005) Antibacterial activities of peptides from the water-soluble extracts of Italian cheese varieties. J Dairy Sci 88(7):2348–2360

Robinson PS, Streusand JV, Chatfield JM et al (1988) Purification and assay of RuBisCo activase from leaves. Plant Physiol 88:1008–1014

Satoh M, Minami M (1995) Molecular pharmacology of the opioid receptors. Pharm Ther 68:343–364

Schmelzer CE, Schöps R, Ulbrich-Hofmann R et al (2004) Mass spectrometric characterization of peptides derived by peptic cleavage of bovine β-casein. J Chromatogr A 1055(1):87–92

Schmelzer CE, Schöps R, Ulbrich-Hofmann R et al (2007) Peptic digestion of β-casein: time course and fate of possible bioactive peptides. J Chromatogr A 1166(1-2):108–115

Sharma SK, Klee WA, Nirenberg M (1975) Dual regulation of adenylate cyclase accounts for narcotic dependence and tolerance. Proc Natl Acad Sci 72(8):3092–3096

Tatham AS, Gilbert SM, Fido RJ et al (2000) Extraction, separation, and purification of wheat gluten proteins and related proteins of Barley, Rye, and Oats. Methods Mol Med 41:55–73

Teschemacher H (2003) Opioid receptor ligands derived from food proteins. Curr Pharm Des 9:1331–1344

Toelstede S, Hofmann T (2008) Sensomics mapping and identification of the key bitter metabolites in Gouda cheese. J Agric Food Chem 56(8):2795–2804

Udenigwe CC (2014) Bioinformatics approaches, prospects and challenges of food bioactive peptide research. Trends Food Sci Technol 36(2):137–143

Udenigwe CC, Gong M, Wu S (2013) In silico analysis of the large and small subunits of cereal RuBisCo as precursors of cryptic bioactive peptides. Process Biochem 48(11):1794–1799

Van Berkel PHC, Geerts MEJ, Van Veen HA et al (1995) Glycosylated and unglycosylated human lactoferrins both bind iron and show identical affinities towards human lysozyme and bacterial lipopolysaccharide but differ in their susceptibilities towards tryptic proteolysis. Biochem J 312:107–114

Wang C, Li D, Xu F et al (2014) Comparison of two methods for the extraction of fractionated rice bran protein. J Chem 51(9):816–827

Zioudrou C, Streaty RA, Klee WA (1979) Opioid peptides derived from food proteins. J Biol Chem 254(7):2446–2449

Zittle CA, Custer JH (1963) Purification and some of the properties of alpha-casein and kappa-casein. J Dairy Sci 46:1183–1188

Conclusions and Future Perspectives 9

Abstract

Food-derived exorphins are usually 4–20 amino acid peptides released from food proteins like caseins, whey, gluten, RuBisCo, β-conglycinin, and albumin. They are generated through gastrointestinal digestion, simulated gastrointestinal digestion (SGID), or fermentation. They show structural features that enable them to bind opioid receptors (μ, δ, and κ). These opioid receptors are distributed widely in the central and peripheral nervous system, gastrointestinal tract, some immune cells, and other tissues. These exorphins may act as agonists or antagonists for these opioid receptors. They play various physiological roles in gastrointestinal motility, analgesia, anxiolysis, emotional and behavior development, memory consolidation, blood pressure regulation, prolactin secretion, food and fat intake, hormone release, appetite, and mucous formation. Moreover, these exorphins are correlated with the development of various physiological complications. The conventional opioids show numerous side effects and thereby limit the clinical effectiveness. Food-derived exorphins are safe alternatives for the pharmaceutical and food industry. The merits include oral consumption and bioavailability, safety due to a lack of side effects, and activation at the endogenous opioid receptors. Therefore, these peptides have enough scope in the food and pharmaceutical industry for the development of functional foods and nutraceuticals. Moreover, deeper research with explored signal cascade mechanisms at the cellular and molecular level is needed to explore food exorphins as therapeutic mediators, functional foods, or nutraceuticals for human health promotion.

9.1   Structural Features of Food-derived Exorphins

Food exorphins are produced from milk, wheat, rice, soybean, spinach, etc. These may be released on gastrointestinal digestion, fermentation, or enzymatic hydrolysis. They mimic the endogenous opioids in structure and potential binding to

M. R. Ul Haq, *Opioid Food Peptides*, https://doi.org/10.1007/978-981-15-6102-3_9

endogenous opioid receptors. These include β-casomorphins (BCMs), casoxins, lactorphins, lactoferroxin, gluten exorphins, soymorphins, and rubiscolin. BCMs bind μ-receptors quite similar to morphine, although affinity is low (Liebmann et al. 1986). The presence of nitrogen atom in the morphine in a well-defined spatial arrangement with respect to the A-aromatic ring confers it the opioid activity. The existence of a phenolic hydroxyl group further helps in strengthening the ligand-receptor affinity. Moreover, the presence of additional F-ring is also significant (Brandt et al. 1994). These critical elements needed for opioid activity are also present in exorphins like BCMs. These structural resemblances pave way for the two different classes of compounds to demonstrate similar opioid activities. Most of the food exorphins have a tyrosine residue at the N-terminal end, followed by an aromatic residue at the 3rd or 4th position (Meisel and FitzGerald 2000). This structural peculiarity enables these exorphins to fit in the binding site of opioid receptors (Teschemacher et al. 1997; Pihlanto-Leppälä 2000). BCMs and lactorphins from milk source may act as opioid agonists, while casoxins from κ-casein serve as opioid antagonists.

BCMs are composed of 4–11 amino acid residues (Kaminski et al. 2007) and share first three residues including tyrosine, proline, and phenylalanine (Muehlenkamp and Warthesen 1996). The demonstration of the opioid activity further depends on the confined negative prospective in the area of the phenolic hydroxyl group of tyrosine, as the elimination of this residue leads to lack of opioid activity (Kostyra et al. 2004). The primary structure of all the bovine BCMs starts from position 60 to 68 (bBCM-8). Human BCMs (BCM-4, 5, 7, and 8) are produced from β-casein of milk. The first three residues (tyrosine, proline, and phenylalanine) from the amino-terminal are shared between human and bovine BCMs. However they differ in the rest of the residues, human BCMs possess a sequence (VEPI) that is different from bovine BCMs (PGPI). The human BCMs are located from 51 to 58 of β-casein of milk. Moreover another six amino acid BCM opioid peptide (neocasomorphin-6) is released with a sequence (YPVEPF) from position 114–119 of bovine β-casein. α-casomorphin is released from the α-casein of human milk (YVPFP) from 158 to 162 of the intact protein.

The whey protein α-lactalbumin from the cow milk yields opioid peptide α-lactorphin on digestion. The opioid peptide is composed of four amino acids, tyrosine at the amino-terminal and phenylalanine at the carboxyl-terminal. The opioid peptide is amidated at the carboxyl-terminal. It is derived from the α-lactalbumin with the fragment 69–72 (Teschemacher et al. 1997; Day et al. 1981). Similarly the cow milk β-lactglobulin on gastrointestinal digestion yield opioid peptide β-lactorphin. It has a sequence (Tyr-Leu-Leu-Phe-NH$_2$). This peptide is composed of four amino acids, tyrosine at the amino-terminal and phenylalanine at the carboxyl-terminal. The opioid peptide is amidated at the carboxyl-terminal (Teschemacher et al. 1997; Day et al. 1981).

Lactoferroxins are milk-derived opioid peptides from lactoferrin. Three types of exorphins are derived from this protein that is denoted by lactoferroxin A, B, and C. All the three are carboxy-methylated at the carboxyl-terminal. Type A is a lactoferrin fragment from 318 to 325 composed of six amino acid residues with a sequence (YLGSGY-OCH$_3$). Type B is a lactoferrin fragment from 536–540 composed of five amino acid residues with a sequence (RYYGY-OCH$_3$). Type C is a lactoferrin fragment from 673–679 composed of seven amino acid residues with a sequence (KYLGPQYOCH$_3$). The peculiarities of all the three opioid peptides are their antagonistic characteristic feature and have µ-opioid receptor selectivity (Tani et al. 1990).

The exorphins from the wheat are either derived from glutenin or alpha-gliadin. The gluten exorphins are derived from glutenin while gliadorphin is released from the alpha-gliadin. A type gluten exorphins possess common four amino acids (GYYP) from N-terminal. B type gluten exorphins share first four amino acids (YGGW) from N-terminal. The gluten exorphin C and gliadorphin share tyrosine at the N-terminal and that is critical for demonstration of opioid activity.

The general structure of the rubiscolin consists of tyrosine at the amino-terminal followed by a proline and leucine. Rubiscolin-5 and 6 have a sequence of YPLDL and YPLDLF, respectively. The general structure of the soymorphins consists of tyrosine at the amino-terminal followed by proline and phenylalanine. All these amino acids are critical for selectivity of the opioid receptors. Soymorphins-5, 6, and 7 have an amino acid sequence of YPFVV, YPFVVN, and YPFVVNA, respectively. Oryzatensin is derived from rice albumin and is composed of nine amino acid residues. The general structure of the oryzatensin consists of GYPMYPLPR. It has a unique structure with glycine at the amino-terminal followed by the tyrosine residue like gluten exorphins A4 and A5.

9.2 Classification of Food-derived Exorphins

BCMs are opioid peptides released from human and bovine α- and β-casein of milk. Based on the number of amino acid residues, bovine and human β-casein generates BCM-4 to BCM-8 besides neocasomorphin-6 from bovine β-casein. Moreover, the human milk releases α-casomorphin from α-casein. Depending upon the source of milk, BCMs may be categorized into bovine and human BCMs.

Lactorphins are classified into two types and generally regarded as alpha (α) and beta (β) forms. α-lactorphin is a four amino acid peptide fragment (69–72) released from α-lactalbumin and amidated at carboxyl-terminal. β-lactorphin is also a four amino acid peptide (118–121) derived from β-lactglobulin and amidated at carboxyl-terminal. Both the types of lactorphins are agonists and have affinity for µ-opioid receptor (Teschemacher et al. 1997; Day et al. 1981).

There are three types of lactoferroxins that are denoted by A, B, and C. These are derived from the lactoferrin of milk protein and represent fragments 318–323, 536–540, and 673–679, respectively. However, compared to other milk peptides these act as antagonist and have affinity for µ-opioid receptors (Tani et al. 1990).

Casoxins are either released from the κ-casein of bovine milk or α-casein of human milk. These exorphins are classified into four types including A, B, C, and D (Yoshikawa et al. 1994).

Based on chain length of peptide and source of proteins, wheat exorphins have six types. Gluten exorphin A4 is derived from glutenin (GYYP) and has far greater selectivity towards δ compared to μ-opioid receptor (Yoshikawa et al. 2003; Fukudome and Yoshikawa 1992). Gluten exorphin A5 is a pentapeptide derived from glutenin (GYYPT) and has far-far greater selectivity towards δ compared to μ-opioid receptor (Yoshikawa et al. 2003; Fukudome and Yoshikawa 1992). Gluten exorphin B4 is generated from glutenin (YGGW) and has selectivity towards δ-opioid receptor (Yoshikawa et al. 2003; Fukudome and Yoshikawa 1992). Gluten exorphin B5 is derived from glutenin (YGGWL) and has selectivity towards δ-opioid receptor (Yoshikawa et al. 2003; Fukudome and Yoshikawa 1992). Gluten exorphin C is derived from glutenin (YPISL) and has selectivity towards delta compare to μ-opioid receptor (Yoshikawa et al. 2003; Fukudome and Yoshikawa 1993). Gliadorphin-7 is a heptapeptide produced from α-gliadin (YPQPQPF) and has selectivity towards delta-opioid receptor.

Rubiscolins are isolated from pepsin hydrolysates of spinach RuBisCo (Yang et al. 2001). There are two types of rubiscolins denoted by rubiscolin-5 and rubiscolin-6. The former has a sequence Tyr-Pro-Leu-Asp-Leu and the latter has a sequence Tyr-Pro-Leu-Asp-Leu-Phe. These opioid peptides have selectivity towards δ-opioid receptor observed through [^{3}H] deltorphin II-binding assay. These also demonstrate opioid activity observed through mouse vas deferens assay (Yang et al. 2001). Rubiscolin-6 shows higher affinity for the opioid receptor (Yang et al. 2003).

Soymorphins are usually chemically synthesized peptides that have a sequence similar to the peptide Tyr-Pro-Phe-Val that is derived from the β-conglycinin β-subunit of soy. The sequence analysis shows that this sequence is similar to the human βh-casomorphin-4. Soymorphins-5, 6, and 7 have sequences Tyr-Pro-Phe-Val-Val, Tyr-Pro-Phe-Val-Val-Asn, and Tyr-Pro-Phe-Val-Val-Asn-Ala, respectively. Soymorphins have selectivity towards μ-opioid receptor. Among these soymorphin-5 demonstrates the highest opioid activity (Ohinata et al. 2007a, b).

9.3   Opioid Activity of Food-Derived Exorphins

The opioid activity of milk-derived exorphins dates back to 1979 when this type of activity was demonstrated for BCMs (Brantl et al. 1979, 1981, 1982; Zioudrou et al. 1979; Koch et al. 1985). Opioid activity was assessed on isolated organ bath on tissues like guinea pig ileum (GPI) and mouse vas deferens (MVD) and ligand-binding assays. Human and bovine BCMs selectively bind MOP, although few BCMs (4 and 5) show affinity towards DOP also (Koch et al. 1985). The order of the potencies are the same for human and bovine BCMs (BCM5 > BCM4 > BCM8 > BCM7). The potency of bovine BCMs is almost 300 times less than morphine. BCM-5 is the most active and the BCM-7 has prolonged effect (Brantl et al. 1981).

The proteolysis (pepsin and trypsin) of bovine κ-casein releases two exorphins with antagonistic properties for their opioid receptors (Yoshikawa et al. 1986; Chiba et al. 1989). One of these exorphins is represented by casoxin C with 10 amino acid residues. The other opioid peptide casoxin with ten amino acid residues with modified C-terminal Tyr residue. Seemingly, the latter gets methoxylated through any of the purification steps. The opioid peptide casoxin (1–6) demonstrated the antagonistic activity for μ-receptors on GPI preparation and the activity of a κ-agonist on the preparation made from rabbit vas deferens. When human α- or κ-casein was digested with pepsin and chymotrypsin (Johnsen et al. 1994), a peptide was purified from α-casein with opioid antagonistic activity (Yoshikawa et al. 1994). This peptide was derived from human αs1 casein from 158-164 and named casoxin D (1–7). This peptide demonstrated very low affinity towards the receptor and antagonized μ-agonist in the preparations made from guinea pig ileum in addition to the inhibitory activity of a δ agonist in the preparation made from mouse vas deferens. It was also reported that all showed antagonistic selectivity almost two orders of magnitude lower compared to the naloxone. The casoxin (1–6) is considered as the μ- and κ-receptor antagonist with low affinity. The selectivities and opioid potencies for synthetic analogs and casoxin D seem to be in range. Nevertheless, the fragments from k-casein demonstrated negligible activities. Earlier Yoshikawa and coworkers in 1986 isolated a bioactive peptide with antagonist opioid activity. The amino acid sequence of the peptide was Ser-Arg-Tyr-Pro-Ser-TyrOCH$_3$ and the peptide was regarded as casoxin 4. The bioactive peptide is a fragment from 33 to 38 of κ-casein with carboxyl ester at C-terminal. The isolated peptide showed activity (IC$_{50}$ = 1 nM) in presence of naloxone (15 mM) when observed through radioreceptor assay. Casoxin C, a physiologically active peptide was isolated from the kappa-casein hydrolysate that was digested with trypsin. It was found to be an anti-opioid bioactive peptide when its activity was subjected to guinea pig ileum (GPI). Results have further established that this peptide demonstrated contractions in the ileal strips (Yoshikawa et al. 1986).

Fukudome and Yoshikawa in 1992 isolated four opioid exorphins from the wheat gluten. The peptides were named as gluten exorphin A5 (Gly-Tyr-Tyr-Pro-Thr), gluten exorphin A4 (Gly-Tyr-Tyr-Pro), gluten exorphin B5 (Tyr-Gly-Gly-Trp-Leu), and gluten exorphin B4 (Tyr-Gly-Gly-Trp). These experiments of assessment of opioid activity were performed in presence of L-leucyl-L-leucine (2 mM) together with bestatin (30 μM), thiorphun (0.3 μM), and captopril (10 μM). The IC$_{50}$ values are the concentration of these opioid peptides in the hydrolysates that inhibits the stimulated (electrically) muscle contraction by 50%. The assessment of opioid activities of these peptides through the MVD assay was done in presence and absence of naloxone (10^{-6} μM). The opioid activities that were reversed on addition of naloxone and blocked by pretreatment with this compound were regarded as opioid activities. The radioreceptor assays were performed as per Pert and Snyder (1973), for κ-receptors in presence of [^{3}H] EKC (1 nM, 24.5 Ci/mmol) and through guinea pig cerebellum membrane. These opioid activity determination assays were performed in presence of inhibitors and this exorphin was found to show high selectivity towards DOP. Gluten exorphin B4 showed the presence of [Trp^4, Leus^5] enkephalin and is the most potent among all purified exorphins with IC$_{50}$ values 0.05 μM and 0.017 μM on GPI and MVD tissues, respectively (Fukudome and Yoshikawa 1992). A novel exorphin was released from wheat with a sequence Tyr-Pro-Ile-Ser-Leu after enzymatic hydrolysis. The IC$_{50}$ values determined were

40 μM and 13.5 μM when observed on the GPI and MVD tissues, respectively. The purified food-derived peptide was denoted as gluten exorphin C (Fukudome and Yoshikawa 1993).

Rubiscolins (5 and 6) are selective for δ-opioid receptors with IC_{50} of 2.09 and 0.093 μM, respectively. They further demonstrate antinociceptive activity (Yang et al. 2001). Comparatively, soymorphins (5 and 7) demonstrate lower IC_{50} on GPI compared to MVD assay. They are selective for μ-opioid receptors and show anxiolytic properties (Ohinata et al. 2007a, b). Oryzatensin, an opioid antagonist for μ-receptors causes biphasic contractions on ileum with a rapid followed by slower contraction. Application of inhibitors (atropine and tetrodotoxin) demonstrated slower contraction mediation through cholinergic nervous system (Takahashi et al. 1994).

9.4 Production of Food-Derived Exorphins

Food exorphins are released from the food proteins on gastrointestinal digestion, SGID, or fermentation. These are usually 4–20 amino acid peptides that have affinity for various opioid receptors and demonstrate numerous biological effects (Korhonen and Pihlanto 2006; Udenigwe et al. 2012; Choi et al. 2012).

Many of the food products are manufactured through the fermentation process including sauerkraut, yogurt, salami, kimchi, bread, and beverages. Fermentation employs use of microorganisms that secrete proteases to cleave proteins to either individual amino acid or peptides. These latter show many properties and among them is a class of peptides that express opioid activity. BCM immunoreactive materials were recovered from fermented milk (*Bacillus cereus* or *Pseudomonas aeruginosa*) (Hamel et al. 1985). Pretreatment of the fermented milk (*Lactobacillus GG*) with the proteolytic system also released opioid peptides (Rokka et al. 1997). Fermented whey with the *Kluyveromyces marxianus var. marxianus* showed the presence of YLLF (β-lactorphin) released from β-lactglobulin (Belem et al. 1999). DPPIV deficient bacterial strains on fermentation of milk generated BCM-4 (Matar and Goulet 1996). BCM precursors were detected when yogurt was fermented with *S. salivarius ssp. Thermophiles* and *L. delbrueckii ssp. Bulgaricus* (Schieber and Brückner 2000). Jarmołowska and coworkers for the first time detected BCM-7 in Brie cheese and latter other investigators also estimated the presence of BCM-7 (6.48 to 0.15 μg/g) (Sienkiewicz-Szłapka et al. 2009; De Noni and Cattaneo 2010). BCM-7 ranges from 0.01 to 0.11 μg/g in a number of cheese samples including Cheddar cheeses, Gorgonzola, Gouda, and Fontina. The SGID treatment further increased the yield up to 21.77 and 15.22 μg/g in Gouda and Cheddar cheeses, respectively (De Noni and Cattaneo 2010). Moreover, BCMs, casoxins, and lactoferroxin A were identified in Gouda, Edamski, and Kasztelan cheeses (Sienkiewicz-Szłapka et al. 2009). The generations of BCMs after SGID by various enzymes have been analyzed in milk and other dairy products (De Noni and Cattaneo 2010) and differential release has further been elucidated in cow β-casein variants (Cieslinska et al. 2007; De Noni 2008; Raies et al. 2015).

Fukudome and Yoshikawa in 1992 detected release of gluten exorphins A4, A5, B4, and B5 in hydrolysates digested with pepsin-thermolysin (Fukudome and Yoshikawa 1992). Relatively, the production of gluten exorphin A5 dominated in pepsin-elastase (250 µg/g) compared o pepsin-thermolysin (40 µg/g) hydrolysates (Fukudome et al. 1997). Gluten exorphin A5 was generated from gluten (0.747–2.192 mg/kg) and exorphin C (3.201–6.689 mg/kg) from bread and pasta after SGID with pepsin-trypsin-chymotrypsin) (Stuknytė et al. 2015). Kong and coworkers checked the opioid activity in pepsin (24 h) and alcalase (6 h) hydrolysates. The IC_{50} values were 1.21 and 1.57 mg/ml for alcalase and pepsin hydrolysates, respectively (Kong et al. 2008).

Bioactive peptides were separated from lactoferrin of milk through enzymatic hydrolysis. The analysis was performed on RP-HPLC and the peptides were generated from the segments 318–323, 536–540, and 673–679. Mass spectroscopy analysis revealed the sequences as Tyr-Leu-Gly-Ser-Gly-Tyr-OCH_3, Arg-Tyr-Tyr-Gly-Tyr-OCH_2, and Lys-Tyr-Leu-Gly-Pro-Gln-Tyr-OCH_3. These opioid peptides were named as lactoferroxin A, B, and C (Tani et al. 1990). Pepsin digestion of κ-casein showed the release of casoxin 4. RP-HPLC and sequence analysis revealed the amino acid sequence of this peptide (Ser-Arg-Tyr-Pro-Ser-TyrOCH_3) (Yoshikawa et al. 1986). The other investigators used the same hydrolysis protocols and sequencing methods for the isolation of casoxins from κ-casein of bovine milk. The results indicated a bioactive peptide fragment corresponding to 25–34 of bovine κ-casein with a sequence Tyr-Ile-Pro-Ile-Gln-Tyr-Val-Leu-Ser-Arg. The peptide was regarded as casoxin C and showed activity at 5 µM through GPI assay (Chiba et al. 1989).

Rubiscolin-5 was produced from spinach RuBisCo through pepsin hydrolysis and the addition of LAP further increased the yield of this exorphin to 5.0%. However, these conditions could not generate rubiscolin-6. The synthesized 14 amino acid residue precursor corresponding to 98–111 of spinach RuBisCo large subunit after pepsin hydrolysis released both rubiscolin-5 (YPLDL, 0.30%) and rubiscolin-6 (YPLDLF, 0.04%) (Yang et al. 2001). Ohinata and coworkers took synthetic peptide and β-conglycinin and hydrolyzed with elastase and LAP. The results indicated release of soymorphin-5 from the synthetic peptide and soy β-conglycinin with 36.5 and 9.1 mol %, respectively. Moreover, it was found that soymorphins-6 and -7 are not produced form either of the two aforesaid sources (Ohinata et al. 2007a, b). Takahashi and coworkers isolated oryzatensin opioid peptide from rice seeds *(Oryza sativa L. var. japonica cv. Nipponbare)* after trypsin digestion *(E/S = 1/100)* at 37 °C for 5 h. The GPI assay demonstrated the opioid antagonistic activity of the fractions of tryptic hydrolysates of rice albumin. A novel peptide named oryzatensin was isolated with a sequence GYPMYPLPR (Takahashi et al. 1994).

9.5   Biological Activities of Food-Derived Exorphins

Lactorphins improve the vascular relaxation in adult spontaneously hypertensive rats (SHR) in vitro. The beneficial role of the alpha-lactorphin was engaged in the direction of endothelial function, nevertheless, beta-lactorphin increased the relaxation that was endothelium-independent (Sipola et al. 2002). The chemically synthesized exorphin with a sequence similar to bovine alpha-lactorphin (Tyr-Gly-Leu-Phe) has a potential significance in the maintenance of the cardiovascular function (Nurminen et al. 2000). Alpha and beta-lactorphins and beta-lactotensin showed ACE inhibitory activities (Mullally et al. 1997) while beta-lactorphin additionally demonstrated osteoprotective potential through decreasing inflammatory and increasing bone formation markers (Pandey et al. 2018). Moreover, α-lactorphin showed enough potential in up-regulating the expression of mucin gene and in turn maintenance of mucosal immunity (Martínez-Maqueda et al. 2012).

Casoxin 4 reverses the inhibitory action in the contractions induced by morphine in guinea pigs and mice in vitro. However, it is unable to affect the morphine inhibitory activity in mouse small intestinal transit through oral administration (Patten et al. 2011). Moreover it may be beneficial in ameliorating the conditions like autoimmune and infectious diseases in addition to the lung and cardiovascular ailments. Therefore, this exorphin has a tremendous scope in the pharmacological industry for preparation of formulations especially in liquid buffer or lyophilisate. These formulations may be provided through mother's milk supplemented with this peptide at doses greater than already present lyoprotectant, cryoprotectant, and excipient (Dorian 2009).

Many correlations have associated autistic and schizophrenic patients with increased absorption of gluten exorphins. Therefore, these peptides may have links with these health complications (Cade et al. 2000). A possible role of gluten exorphin in pituitary function has been demonstrated and that too occurs through conventional opioid receptors (Fanciulli et al. 2002). Gluten exorphin B5 secretes prolactin when adminstered through peripheral route and this regulation occurs via opioid receptor pathway. However, the dose of the peptide given will have a possible function in the secretion of prolactin (Fanciulli et al. 2003). Gluten exorphin B4 compared to earlier studied (B5) could not alter the secretion of prolactin in rats, although this peptide was given at doses three times higher compared to earlier analyzed peptide. The authors finally concluded that leucine at the carboxyl end is significant for its function in secreting prolactin (Fanciulli et al. 2004). Gluten exorphin 5 has a potential role in modification of the brain neurotransmitter release without transporting through the blood–brain barrier (BBB) (Fanciulli et al. 2005). Gluten exorphin A5 produces diverse effects in the central as well as peripheral nervous systems (Takahashi et al. 2000). Although many reports have established the role of endorphins in the modulation of appetite, nevertheless, these investigators could not observe any significant effect on the quantity of calories taken or satiety perception. Therefore, the results demonstrate a potential impact of these exorphins from various diets in the management of intestinal function (Morley et al. 1983)

Gluten hydrolysate assumed to have exorphins that stimulates the secretion of postprandial glucagon and insulin (Schusdziarra et al. 1981).

Rubiscolins enhance glucose uptake possibly through AMPK activation to increase GLUT4 translocation after binding opioid receptors (Kairupan et al. 2019). It is suggested that this peptide induced antidepressant-like activity mediated through δ-opioid receptors (Mitsumoto et al. 2019). The oral bioavailability of this exorphin and activation of the δ-opioid receptors regards it as potential drug against alcohol use disorder (Chiang et al. 2016; Robins et al. 2018) and an analgesic with lesser side effects. Rubiscolin-6 encourages food intake in the aged mice with an intact ghrelin resistance (Miyazaki et al. 2014). Rubiscolins decrease consumption of high fat diet possibly through activation of central δ-opioid receptor system (Kaneko et al. 2014). This exorphin may be a potential exogenous complement that may aid the endogenous defense mechanisms of humans under stress conditions (Artemova et al. 2010) and helps in the regulation (suppression) of the protective behavior in insects (Gritsaĭ et al. 2009). Rubiscolins have remarkable and prolonged effects on the adaptation (environmental) of young mammals (Dubynin et al. 2008). Rubiscolin-mediated anxiolytic effect occurs through sigma and dopamine (1) receptors downstream of δ-opioid receptor (Hirata et al. 2007).

Soymorphins induce anxiolysis at doses different that are needed to induce analgesis (Ohinata et al. 2007a, b). These observations therefore advocate that 5-HT (1A) activation produces D (2) proceeded by GABA (B) activation in intestinal motility. Therefore, soymorphins on oral administration decrease the food intake and intestinal transit through M1OP coupled to GABA (B), 5-HT (1A), and D (2) systems (Kaneko et al. 2010).

Oryzatensin increased the expression of VEGF and MMP-9 in HEP-G2 cell lines. Therefore, this peptide has a potential role in immunomodulation through sensitization of cancer cells (Barati et al. 2017). Also, this peptide was found to improve amnesia induced by scopolamine and ischemia (Jinsmaa et al. 2001). Therefore this peptide has a possible role in suppression of food intake possibly through PGE2 generation and EP4 receptor activation (Ohinata et al. 2007a, b).

9.6 Overall Conclusions

Food exorphins are usually 4–20 amino acid peptides released from the intact proteins through gastrointestinal digestion, fermentation, and proteolysis. These exorphins are generated from food proteins like milk, wheat, soybean, rice, and spinach. They show structural features that enable them to bind the endogenous opioid receptors. Opioid receptors (μ, δ, and κ) are distributed widely in the central and peripheral nervous system, gastrointestinal tract, and some immune cells. These exorphins may act as agonist or antagonist for these opioid receptors. They play various physiological roles in gastrointestinal motility, analgesis, anxiolysis, emotional and behavior development, blood pressure regulation, hormone release, appetite, and mucous formation. Moreover, these peptides have been correlated with the development of various physiological complications including type 1 diabetes, cardiovascular diseases, sudden infant death syndrome, celiac disease, neurological manifestation (autism and schizophrenia).

9.7   Future Perspectives

The conventional opioids like alkaloids (morphine) demonstrate numerous side effects and thereby limit the clinical effectiveness. On the other hand, the food-derived exorphins provide alternate options for pharmaceutical and food industry. These exorphins provide some merits over conventional opioids that include oral consumption and bioavailability, safety due to lack of side effects, and activation at the endogenous opioid receptors. Therefore, these peptides have enough scope in the food and pharmaceutical industry for the development of functional foods and nutraceuticals. They have ample potential in regulating neurological, gastrointestinal, and cardiovascular functions. However, the opioid peptides that are linked with few human illnesses should not be encouraged in the food and pharmaceutical industry. Moreover, deeper research with explored signal cascade mechanism at cellular and molecular level is needed to explore food exorphins as therapeutic mediators, functional foods, or nutraceuticals for human health promotion.

References

Artemova NV, Bumagina ZM, Kasakov AS et al (2010) Opioid peptides derived from food proteins suppress aggregation and promote reactivation of partly unfolded stressed proteins. Peptides 31 (2):332–338

Barati M, Yousefi M, Ebrahimi-Mameghani M (2017) Oryzatensin-stimulated PBMCs increase cancer progression in-vitro. Iran J Allergy Asthm 16(2):120–126

Belem MA, Gibbs BF, Lee BH et al (1999) Proposing sequences for peptides derived from whey fermentation with potential bioactive sites. J Dairy Sci 82(3):486–493

Brandt W, Barth A, Höltje HD (1994) Investigations of structure-activity relationships of β-casomorphins and further opioids using methods of molecular graphics. In: Brantl V, Teschemacher H (eds) β-casomorphins and related peptides: recent developments. Wiley, Weinheim, pp 93–104

Brantl V, Teschemacher H, Henschen A et al (1979) Novel opioid peptides derived from casein (beta-casomorphins). I. Isolation from bovine casein peptone. Hoppe Seylers Z Physiol Chem 360(9):1211–1216

Brantl V, Teschemacher H, Blasig J et al (1981) Opioid activities of beta-casomorphins. Life Sci 28 (17):1903–1909

Brantl V, Pfeiffer A, Herz A et al (1982) Antinociceptive potencies of beta-casomorphin analogs as compared to their affinities towards mu and delta opiate receptor sites in brain and periphery. Peptides 3(5):793–797

Cade R, Privette R, Fregly M et al (2000) Autism and schizophrenia: intestinal disorders. Nutr Neurosci 3:57–72

Chiang T, Sansuk K, Van Rijn RM (2016) Beta-arrestins 2 dependence of delta opioid receptor agonists is correlated with alcohol intake. Br J Pharmacol 173:323–343

Chiba H, Tani F, Yoshikawa M (1989) Opioid antagonist peptides derived from κ-casein. J Dairy Res 56:363–366

Choi J, Sabikhi L, Hassan A, Anand S (2012) Bioactive peptides in dairy products. Int J Dairy Technol 65(1):1–12

Cieslinska A, Kaminski S, Kostyra E et al (2007) β-Casomorphin-7 in raw and hydrolyzed milk derived from cows of alternative β-casein genotypes. Milchwissenschaft 62:125–127

Day A, Freer R, Liao C (1981) Morphiceptin (beta-casomorphin (1-4) amide): a peptide opioid antagonist in the field stimulated rat vas deferens. Res Commun Chem Pharmacol 34 (3):543–546

De Noni I (2008) Release of beta-casomorphins 5 and 7 during simulated gastro-intestinal digestion of bovine beta-casein variants and milk-based infant formulas. Food Chem 110(4):897–903

De Noni I, Cattaneo S (2010) Occurrence of b-casomorphins 5 and 7 in commercial dairy products and in their digests following in vitro simulated gastro-intestinal digestion. Food Chem 119:560–566

Dorian B (2009) Use of a peptide as a therapeutic agent. PCT/EP2008/007842

Dubynin VA, Malinovskaya IV, Belyaeva YA (2008) Delayed effect of exorphins on learning of albino rat pups. Biol Bull 35:43–49

Fanciulli G, Dettori A, Tomasi PA et al (2002) Prolactin and growth hormone response to intracerebroventricular administration of the food opioid peptide gluten exorphin B5 in rats. Life Sci 71(20):2383–2390

Fanciulli G, Dettori A, Fenude E et al (2003) Intravenous administration of the food-derived opioid peptide gluten exorphin B5 stimulates prolactin secretion in rats. Pharmacol Res 47(1):53–58

Fanciulli G, Dettori A, Demontis MP et al (2004) Gluten exorphin B5 stimulates prolactin secretion through opioid receptors located outside the blood-brain barrier. Life Sci 76(15):1713–1719

Fanciulli G, Dettori A, Demontis MP et al (2005) Serum prolactin levels after administration of the alimentary opioid peptide gluten exorphin B4 in male rats. Nutr Neurosci 7(1):53–55

Fukudome S, Yoshikawa M (1992) Opioid peptides derived from wheat gluten: their isolation and characterization. FEBS Lett 296(1):107–111

Fukudome S, Yoshikawa M (1993) Gluten exorphin C. A novel opioid peptide derived from wheat gluten. FEBS Lett 316(1):17–19

Fukudome S, Jinsmaa Y, Matsukawa T (1997) Release of opioid peptides, gluten exorphins by the action of pancreatic elastase. FEBS Lett 412(3):475–479

Gritsaĭ OB, Dubynin VA, Bespalova ZD et al (2009) Effects of several exorphins and endorphins on the escape reaction of the cockroach Periplaneta Americana under elevated temperature conditions. Zh Evol Biokhim Fiziol 45(4):391–407

Hamel U, Kielwein G, Teschemacher H (1985) Beta-casomorphin immunoreactive materials in cows' milk incubated with various bacterial species. J Dairy Res 52(1):139–148

Hirata H, Sonoda S, Agui S et al (2007) Rubiscolin-6, a δ opioid peptide derive from spinach RuBisCo, has anxiolytic effect via activating σ1 and dopamine D1 receptors. Peptides 28 (10):1998–2003

Jinsmaa Y, Takenaka Y, Yoshikawa M (2001) Designing of an orally active complement C3a agonist peptide with anti-analgesic and anti-amnesic activity. Peptides 22(1):25–32

Johnsen LB, Rasmussen LK, Petersen TE et al (1994) Characterization of three types of human alpha s1-casein mRNA transcripts. Biochem J 309:237–242

Kairupan TS, Cheng KC, Asakawa A (2019) Rubiscolin-6 activates opioid receptors to enhance glucose uptake in skeletal muscle. J Food Drug Anal 27(1):266–274

Kaminski S, Cieoelinska A, Kostyra E (2007) Polymorphism of bovine beta-casein and its potential effect on human health. J Appl Genet 48(3):189–198

Kaneko K, Iwasaki M, Yoshikawa M et al (2010) Orally administered soymorphins, soy-derived opioid peptides, suppress feeding and intestinal transit via gut μ1-receptor coupled to 5-HT1A, D2, and GABAB systems. Am J Physiol Gastrointest Liver Physiol 299(3):799–805

Kaneko K, Mizushige T, Miyazaki Y et al (2014) δ-Opioid receptor activation stimulates normal diet intake but conversely suppresses high-fat diet intake in mice. Am J Physiol Regul Integr Comp Physiol 306(4):265–272

Koch G, Wiedemann K, Teschemacher H (1985) Opioid activities of human β-casomorphins. Arch Pharmacol 331(4):351–354

Kong X, Zhou H, Hua Y et al (2008) Preparation of wheat gluten hydrolysates with high opioid activity. Eur Food Res Technol 227(2):511–517

Korhonen H, Pihlanto A (2006) Bioactive peptides: production and functionality. Int Dairy J 16 (9):945–960

Kostyra E, Sienkiewicz-Szapka E, Jarmolowska B et al (2004) Opioid peptides derived from milk proteins. Pol J Food Nutr 13(54):25–35

Liebmann C, Barth A, Neubert K et al (1986) Effects of β-Casomorphin on 3h-ouabain binding to guinea-pig heart membranes. Pharamzie 41:670–671

Martínez-Maqueda D, Miralles B, De Pascual-Teresa S et al (2012) Food-derived peptides stimulate mucin secretion and gene expression in intestinal cells. J Agric Food Chem 60(35):8600–8605

Matar C, Goulet J (1996) β-casomorphin 4 from milk fermented by a mutant of Lactobacillus helveticus. Int Dairy J 6(4):383–397

Meisel H, FitzGerald RJ (2000) Opioid peptides encrypted in intact milk protein sequences. Br J Nutr 84(1):27–31

Mitsumoto Y, Sato R, Tagawa N et al (2019) Rubiscolin-6, a δ-opioid peptide from spinach rubisco, exerts antidepressant-like effect in restraint-stressed mice. J Nutr Sci Vitaminol 65(2):202–204

Miyazaki Y, Kaneko K, Iguchi S (2014) Orally administered δ opioid agonist peptide rubiscolin-6 stimulates food intake in aged mice with ghrelin resistance. Mol Nutr Food Res 58 (10):2046–2052

Morley JE, Levine AS, Yamada T et al (1983) Effect of exorphins on gastrointestinal function, hormonal release, and appetite. Gastroenterology 84(6):1517–1523

Muehlenkamp MR, Warthesen JJ (1996) Beta-casomorphins: analysis in cheese and susceptibility to proteolytic enzymes from *Lactococcus lactis ssp Cremoris*. J Dairy Sci 79(1):20–26

Mullally MM, Meisel H, FitzGerald RJ (1997) Identification of a novel angiotensin-I-converting enzyme inhibitory peptide corresponding to a tryptic fragment of bovine b-lactoglobulin. FEBS Lett 402:99–101

Nurminen ML, Sipola M, Kaarto H et al (2000) α-Lactorphin lowers blood pressure via radiotelemetry in normotensive and spontaneously hypertensive rats. Life Sci 66:1535–1543

Ohinata K, Agui S, Yoshikawa M (2007a) Soymorphins, novel μ opioid peptides derived from soy β-conglycinin β-subunit, have anxiolytic activities. Biosci Biotechnol Biochem 71 (10):2618–2621

Ohinata K, Agui S, Yoshikawa M (2007b) Soymorphins, novel μ opioid peptides derived from soy β-conglycinin β-subunit, have anxiolytic activities. Biosci Biotechnol Biochem 71 (10):2618–2621

Pandey M, Kapila S, Kapila R et al (2018) Evaluation of the osteoprotective potential of whey derived-antioxidative (YVEEL) and angiotensin-converting enzyme inhibitory (YLLF) bioactive peptides in ovariectomised rats. Food Funct 9(9):4791–4801

Patten GS, Head RJ, Abeywardena MY (2011) Effects of casoxin 4 on morphine inhibition of small animal intestinal contractility and gut transit in the mouse. Clin Exp Gastroenterol 4:23–31

Pert CB, Snyder SH (1973) Opiate receptor: demonstration in nervous tissue. Science 179 (4077):1011–1014

Pihlanto-Leppälä A (2000) Bioactive peptides derived from bovine whey proteins. Trends Food Sci Technol 11:347–356

Raies MH, Kapila R, Kapila S (2015) Release of β-casomorphin-7/5 during simulated gastrointestinal digestion of milk β-casein variants from Indian crossbred cattle (Karan fries). Food Chem 1 (168):70–79

Robins MT, Chiang T, Mores KL et al (2018) Critical role for Gi/o-Protein activity in the dorsal striatum in the reduction of voluntary alcohol intake in C57Bl/6 Mice. Front Psych 9:112

Rokka T, Eeva-Liisa S, Jari T et al (1997) Release of bioactive peptides by enzymatic proteolysis of Lactobacillus GG fermented UHT milk. Milchwissenschaft 52:675–678

Schieber A, Brückner H (2000) Characterization of oligo-and polypeptides isolated from yoghurt. Eur Food Res Technol 210(5):310–313

Schusdziarra V, Henrichs I, Holland A (1981) Evidence for an effect of exorphins on plasma insulin and glucagon levels in dogs. Diabetes 4:362–364

Sienkiewicz-Szłapka E, Jarmołowska B, Krawczuk S et al (2009) Contents of agonistic and antagonistic opioid peptides in different cheese varieties. Int Dairy J 19(4):258–263

Sipola M, Finckenberg P, Korpela R et al (2002) Effect of long-term intake of milk products on blood pressure in hypertensive rats. J Dairy Res 69:103–111

Stuknytė M, Maggioni M, Cattaneo S et al (2015) Release of wheat gluten exorphins A5 and C5 during in vitro gastrointestinal digestion of bread and pasta and their absorption through an in vitro model of intestinal epithelium. Food Res Int 72:208–214

Takahashi M, Moriguchi S, Yoshikawa M et al (1994) Isolation and characterization of oryzatensin: a novel bioactive peptide with ileum-contracting and immunomodulating activities derived from rice albumin. Biochem Mol Biol Int 33(6):1151–1158

Takahashi M, Fukunaga H, Kaneto H et al (2000) Behavioral and pharmacological studies on gluten exorphin A5, a newly isolated bioactive food protein fragment, in mice. Jpn J Pharmacol 84 (3):259–265

Tani F, Iio K, Chiba H et al (1990) Isolation and characterization of opioid antagonist peptides derived from human lactoferrin. Agric Biol Chem 54(7):1803–1810

Teschemacher H, Koch G, Brantl V (1997) Milk protein-derived opioid receptor ligands. Biopolymers 43(2):99–117

Udenigwe CC, Adebiyi AP, Doyen A et al (2012) Low molecular weight flaxseed protein-derived arginine-containing peptides reduced blood pressure of spontaneously hypertensive rats faster than amino acid form of arginine and native flaxseed protein. Food Chem 132:468–475

Yang S, Yunden J, Sonoda S et al (2001) Rubiscolin, a δ selective opioid peptide derived from plant RuBisCo. FEBS Lett 509(2):213–217

Yang S, Kawamura Y, Yoshikawa M (2003) Effect of rubiscolin, a δ opioid peptide derived from RuBisCo, on memory consolidation. Peptides 24(2):325–328

Yoshikawa M, Tani F, Ashikaga T et al (1986) Purification and characterization of an opioid antagonist from a peptide digest of bovine κ-casein. Agric Biol Chem 50:2951–2954

Yoshikawa M, Tani F, Shiota H et al (1994) Casoxin D, an opioid antagonist ileum-contracting/ vasorelaxing peptide derived from human αs1-casein. In: Brantl V, Teschemacher H (eds) β-Casomorphins and related peptides: recent developments. Wiley, Weinheim, pp 43–48

Yoshikawa M, Takahashi M, Yang S (2003) Delta opioid peptides derived from plant proteins. Curr Pharm Des 9(16):1325–1330

Zioudrou C, Streaty RA, Klee WA (1979) Opioid peptides derived from food proteins. J Biol Chem 254(7):2446–2449

The manufacturer's authorised representative in the EU is Springer Nature Customer Service Centre GmbH, Europaplatz 3, 69115 Heidelberg, Germany. If you have any concerns regarding our products, please contact ProductSafety@springernature.com

Printed and bound by CPI Group (UK) Ltd, Croydon, CR0 4YY

01/12/2025

02008619-0008